ruth weiss: Beat Poetry, Jazz, Art

American Frictions

Editors
Carsten Junker
Julia Roth
Darieck Scott

Editorial Board
Arjun Appadurai, New York University
Mita Banerjee, University of Mainz
Tomasz Basiuk, University of Warsaw
Isabel Caldeira, University of Coimbra

Volume 3

Estíbaliz Encarnación-Pinedo, Thomas Antonic
ruth weiss

Beat Poetry, Jazz, Art

DE GRUYTER

ISBN 978-3-11-126218-5
e-ISBN (PDF) 978-3-11-069455-0
e-ISBN (EPUB) 978-3-11-069464-2
ISSN 2698-5349

Library of Congress Control Number: 2021938518

Bibliographic information published by the Deutsche Nationalbibliothek
The Deutsche Nationalbibliothek lists this publication in the Deutsche Nationalbibliografie;
detailed bibliographic data are available on the Internet at http://dnb.dnb.de.

© 2022 Walter de Gruyter GmbH, Berlin/Boston
This volume is text- and page-identical with the hardback published in 2021.
Frontispiece: ruth weiss 1959, © C. R. Snyder
Printing and binding: CPI books GmbH, Leck

www.degruyter.com

Table of Contents

A. Robert Lee
Preface —— XI

Estíbaliz Encarnación-Pinedo, Thomas Antonic
ruth weiss at Last: Introducing the Poet —— XV

Part One: **Beyond Poetry**

Stefan Weber
Tribute to ruth weiss upon receiving the medal of honor by the City of Vienna in 2006 —— 3

Anne Waldman
RUTH IN THE HOUSE —— 7

Tate Swindell
Future A.D. with Tate —— 9

John Wieners
ARCHITECTURE —— 13

Janet DeBar
ruth —— 15

Mary Norbert Körte
THE BIG NINE-OH —— 17

Agneta Falk
RUTHFUL —— 19

S.A. Griffin
can't stop the beat —— 21

Pete Winslow
Soft Memories —— 23

A.D. Winans
Grand Old Diva —— 25

Jack Hirschman
Buon' Compleanno, RUTH!! —— 27

Neeli Cherkovski
In the desert —— 29

Steven Arnold
Day 25 (Diary excerpt) —— 31

Part Two: **Poetry, Jazz & Art**

Benjamin J. Heal
ruth weiss: Transnationalism and Resistance —— 35

Stefanie Pointl
"vienna. not quite.": Place, Movement, and Identity in ruth weiss' Poetry —— 51

Estíbaliz Encarnación-Pinedo
"How real is i?": Gender and Poetics in ruth weiss —— 67

Polina Mackay
ruth weiss and the Poetics of the Desert —— 81

Chad Weidner
Reaching Towards the Light: Transitory Spaces and the Negated Material Body in Selected Texts by ruth weiss —— 99

Hannes Höfer
Traditionally New: The Jazz & Poetry Work of ruth weiss —— 113

Peggy Pacini
"being tested": ruth weiss at the Summer of Love 2007 —— 127

Caroline Crawford
Oral History Interview with ruth weiss —— 147

Frida Forsgren
ruth weiss and Visual Art: The Watercolor Haiku Series *A Fool's Journey* and *Banzai!* —— 159

Lars Movin
"Go to the roundhouse, he can't corner you there": *The Brink* (1961), ruth weiss' Poetic Film —— 179

Steve Seid
ruth weiss, Luminosity Procured —— 195

Thomas Antonic
The ruth weiss Papers —— 205

Thomas Antonic, Paul Pechmann
ruth weiss: Complete Bibliography —— 223

Notes on Contributors —— 259

Index —— 265

A. Robert Lee
Preface

Beat Goddess.

You might think that would be quite sufficient to win anyone's attention.

But once you get to reading ruth weiss, or listening to her pioneer Jazz & Poetry performances, or watching her film and dance work together with her travel stories and haiku and watercolours, it becomes abundantly clear that so much more is signified in her life and art.

You are dealing with latitudes and longitudes. German birth. Vienna to California. Nazi Europe and the shadow of holocaust to West Coast America and 1950s-and-after bohemian North Beach.

En route a whole treasury of sites and residence. Childhood flight to Vienna and The Netherlands. School-time in New York's Harlem, then Chicago and Switzerland. Travels and adventuring to follow in Greenwich Village, New Orleans, and eventually San Francisco. And all the time the poet. To which her more than twenty published volumes, chapbooks and artwork give every confirmation.

Writings in English and in German translation. Verse, to be sure, but also screen, stage, couture, design.

And always the lower-case signing of her name in protest at Germanic *Ordnung* and capitalization of nouns. Add in her signature blue-green dyed hair as environmental gesture, one wholly in keeping with her demise at 92 years of age at the Mendocino County home she shared with her artist partner Paul Blake and later with Hal Davis in Albion amid California's stately redwoods. Add in her jazz (and non-jazz) readings, be it The Cellar, City Lights, The Co-Existence Bagel Shop, Caffe Trieste or each other venue: riff, phrasings, improvisation, sax, flute, percussion.

A life, a modern European-Jewish and Beat-troubadour American life, managed against initial odds but always with huge creative resilience.

A truly singular life lived inside truly singular history.

A Fool's Journey, which came out in a bilingual edition in 2012, she says is "dedicated to all the fools that paint & write & music make from the outside in." It is not hard to recognize both her sense of the elusive challenge of art in general and of her own will to literary fashioning. From the outset her eye and ear were nothing if not pitched to capture the world as kinetic, realms of organic growth and nurture as against history's brute fissures, its too frequent human damage.

Take your further pick. In "Dedication" from *Light and Other Poems* (1976) she has her speaker lay down the desideratum "listen poets/dedicate your lines/ to the circle!" In *Single Out* (1978) the poignant autobiographical title-poem ("at the ps some kids called me nazi/the war was on/I spoke German") and the imagistic "Earth Paintings" speak respectively to her focus in consciousness of history and of the natural world.

These combine in the two volumes and film likely to give her strengths their permanence.

Desert Journal, composed 1966–69 but published a decade later, might almost be a manifesto poem. Its forty-part sequence in its echoes of Moses and the Israelite exodus, Jesus in the desert, and the "Ali Baba and the Forty Thieves" story, offers a reflexive, associational, exploration of the creative mind ("the internal desert"). "The way is always open" gives a Zen-Beat touch, weiss as at once heir to Hebraic and Talmudic tradition and yet a voice committed to her own freely hybrid inflections of vision and image.

Can't Stop the Beat: The Life and Words of a Beat Poet (2011), a quartet of life-diaries in verse and prose with photography, covers wide topographies, among them Vienna, Harlem, South Side Chicago, Africa, the Pacific Ocean, Summer of Love San Francisco, Coit Tower, remembered Mexico with her first husband Mel Weitsman. Here, too, you meet her Beat-Beatnik remembrance, haiku-writing with Jack Kerouac, poetry with Bob Kaufman.

The Brink (1961, video 1986, preservation print by the Pacific Film Archive 2019) remains her film classic, an improvised visual choreography of love and love-pair as free-flow of image. Dance, movement, inscribe a kind of ode to joy against conformity and the rush of conflicting voice. Once known only at Beat margins and underground it has latterly moved more into the light.

Gender of necessity enters the reckoning. If we are getting used to the idea that Beat was not only the Ginsberg, Kerouac and Burroughs trinity, with Corso, Ferlinghetti and Snyder in the frame, then weiss is key among women's Beat authorship. *A Parallel Planet of People and Places: Stories and Poems* (2012) gives generous recognition to her co-makers, forty vignettes no less from Diane di Prima to Jan Kerouac, Joanne Kyger to Janine Pommy Vega. She can even include Billie Holiday and herself in the portraiture ("the beat the beat the beat/the poet to be the instrument").

Some career. Some presence.

To have, then, a first-ever collection of essays given over to her achievement is not only timely but almost spectacularly overdue. Each offers necessary entrance into the cross-lights of weiss's writing and art: Vienna, Desert, Gender, Poetics, Space, Festival, Visuality.

That, too, the collection falls under transnational Spanish-Austrian co-editorship, from a German press, with contributors from the USA, the UK, Austria, Spain, Cyprus, France, Germany, Switzerland, Sweden, Norway and Denmark, and with tributes from Anne Waldman and other countercultural US and European poets, could not be more fitting. weiss was nothing if not transnational in her own terms of place as of imaginative reach.

The documentary made by Melody C. Miller in 2019 as *ruth weiss*, its half-title actually a coining by the legendary San Francisco journalist Herb Caen, attempts one kind of honouring tribute.

Quite another documentary, *ruth weiss: One More Step West Is the Sea,* from the collection's co-editor Thomas Antonic, had its film festival premiere in June 2021.

For their part the essays at hand in *ruth weiss: Beat Poetry, Jazz & Art* give their own timeliest due.

Estíbaliz Encarnación-Pinedo, Thomas Antonic
ruth weiss at Last: Introducing the Poet

In 1952, ruth weiss found herself in San Francisco after hitchhiking through the United States. The future "Goddess of the Beat Generation" – as she was later christened by San Francisco Chronicle critic Herb Caen (a moniker she often detested) – initially moved to an apartment in the heart of North Beach. Doing so much by coincidence rather than design, she ended up living in the same building at 1010 Montgomery where Allen Ginsberg would three years later start writing his landmark poem, "Howl". One of her neighbors in that building happened to be Philip Whalen. It was also two years before she would write the night away composing haikus in the Wentley Hotel with Jack Kerouac, who had yet to publish his best-selling novel *On the Road* (1957). At that time, John Wieners also lived there, writing his famous *Hotel Wentley Poems*. But the Beat Generation phenomenon was still far from weiss, even if she herself had in fact been "on the road" for quite a while.

Born in Berlin in 1928 to Austrian-Jewish parents, the family felt compelled to move back to Vienna to escape the threat of Nazism in 1933, having to flee again after the annexation of Austria into Nazi-Germany in 1938, escaping on the last train permitted to cross the border into the Netherlands. One of the subsequent manifestations of this trauma was the consistent use of all lower-case letters in her name as a symbolic gesture against (any) law and order, especially when it came to her native language. In 1939, weiss and her parents arrived in New York, where she – not yet able to speak English – was put in a children's home in Harlem as the only white girl in her class, thus becoming acquainted of Afro-American culture and slang. In the 1940s the family moved to Chicago where weiss finally started her bohemian life. It was in 1949 when she first read her poetry to live Jazz music – pioneering, for many, the Jazz & Poetry genre.

From that moment onwards, and almost until her death in 2020, weiss performed Jazz & Poetry and worked for almost seven decades in a wide range of artistic forms. She authored over twenty poetry books, wrote over a dozen theatre plays, exhibited her water-color haiku paintings at the San Francisco Museum of Modern Art and other venues, acted in films and even wrote and directed one herself. As such, weiss embodied the artistic confluence of 1950s and 1960s bohemia, and broke down, as Randy Roark writes, "the barriers between word, film, song, painting, and theatre"[1]. And she continued to perform (and later also recorded) Jazz & Poetry over the decades, until the 1990s mainly in California, and later, after gaining more recognition, throughout North America and the whole of Europe. At the age of 90 she was still standing on stage, roaring her poems into the microphone with a deep, smoky voice, accompanied by her jazz trio, for example at San Francisco's Beat Museum. Jack Hirschman, a long-time companion and San Francisco Poet Laureate, commented on weiss' musicality:

> No American poet has remained so faithful to jazz in the construction of poetry as has ruth weiss. Her poems are scores to be sounded with all her riffy ellipses and open-formed phrasing swarming the senses ... Others read to jazz or write from jazz. ruth weiss writes jazz in words.[2]

The fact that she hadn't learned the language in which she wrote poetry until the age of ten also enabled her to take a keener view of her medium, from which she was able to derive astonishing puns and references that are otherwise hard to find in American poetry.

Despite her extensive poetry career and very active participation in the West Coast buzzing artistic community since the early 1950s, weiss has essentially remained an unjustifiably overlooked figure in poetry history. This neglect might be representative – or shall we say a consequence – of the overshadowing of female artists within the Beat Generation as they often represent, as Ronna C. Johnson has noted, "a marginalized group within an always already marginalized bohemia"[3]. This marginalized position has only recently started to be recovered and acknowledged. Indeed, following weiss' re-discovery in the course of Brenda Knight's *Women of the Beat Generation* (1996), a book that included an interview and previously unpublished work, the poet benefitted from a boost in Beat-related academic and cultural activities in the last decades.

Still, even if perhaps more subconsciously than deliberately, academia has retained a blind spot for certain subjects and artists such as ruth weiss. This explains the absence of any reference to weiss in relation to the history of Jazz &

---

[1] Blurb to ruth weiss (2011). *Can't Stop the Beat: The Life and Words of a Beat Poet.* Studio City, CA [Los Angeles]: Divine Arts.
[2] Qtd. in Brenda Knight (1996). *Women of the Beat Generation: The Writers, Artists, and Muses at The Heart of a Revolution.* Berkeley: Conari. Print. Here p. 247.
[3] Ronna C. Johnson (2004). "Mapping Women Writers of the Beat Generation." *Breaking the Rule of Cool: Interviewing and Reading Beat Writers.* Ed. Nancy M. Grace and R. C. J. Jackson: UP of Mississippi. 3–41. [Here p. 5] Print.

Poetry in recent publications on the Beats and music, as well as the mere passing reference to weiss as a playwright in studies on the Beats and drama. Much the same can be said of the now almost systematic omission of ruth weiss' *The Brink* – a film described by no other than the US experimental filmmaker Stan Brakhage as "one of the most ambitious 'first' films [he'd] ever seen"[4] – in volumes on the San Francisco Bay Area film scene. Intended or not, these omissions and half-mentions followed weiss throughout most of her career, assigning her to the ranks of an array of marginal(ized) voices that are continually overlooked.

Such observations, unfortunately, can be made about many other female Beats as well. Truly, twenty-five years after the publication of Knight's groundbreaking anthology and despite the rekindled interest in the movement in general and in the work of women in particular, the Beat Generation academic niche still has tremendous shortcomings in terms of monographs and individual studies dealing with the work of female poets. The present volume, *ruth weiss: Beat Poetry, Jazz and Art*, taps directly into this lacuna by providing up-to-date, comprehensive, critical analyses around one of the most prolific artists of the so-called "Beat women": ruth weiss.

Unearthing ruth weiss' legacy

In light of the above, it's certainly telling that it took until 2021 for the first dedicated volume focusing on the work of a female key figure of the West Coast Beat scene to be published. Truly, this neglect certainly reflects the position of women within a movement that for a long time was seen as predominantly white and masculine. It's only quite recently that this situation has started to change, thanks to the publications of the first single-authored books on women writers associated with the Beat Generation such as Mary P. Carden's *Women Writers of the Beat Era* (2018) and David S. Calonne's *Diane di Prima: Visionary Poetics and the Hidden Religions* (2019). In more ways than one, female poets of the Beat Generation are still trapped in the myth of the "silent chick", pushed to the periphery of the movement and seen – if at all – through the eyes of others. Nevertheless, like Carden's and Calonne's books, *ruth weiss: Beat Poetry, Jazz and Art* does not reclaim or reassert ruth weiss' position amongst the writers of the Beat Generation, nor does it argue that weiss should be studied through the always kaleidoscopic but ultimately limiting Beat lens.

---

4 Stan Brakhage. (1963) "Letter to Jonas Mekas." *Film Culture* 9.29: 80. Print.

Rather, the diverse and comprehensive access points into ruth weiss' oeuvre that are adopted by the essays in this collection stem from weiss' own multifaceted and expansive poetic vision. As such, the volume includes studies on most areas of weiss' body of work – poetry, film, performance and painting – as well as contrastive and comparative analyses of ruth weiss' poetics and aesthetics and that of other poets and artists, both inside and outside the scope of the Beat Generation. In order to respond to the existing cross-pollination between art forms in weiss' oeuvre, this collection maintains a multi-disciplinary approach that is considered a methodological necessity when dealing with weiss' body of work which, in turn, aligns the volume with recent developments in the field. The collection provides detailed analyses of individual collections of poetry while simultaneously analyzing the poet's position as a Jazz & Poetry artist. Furthermore, the volume takes into consideration the various literary genres within which weiss worked (lyric and narrative poem, travel journal, haiku, theatrical play, etc.), as well as her personal biography and its effect on her poetic and artistic vision. Other topics include weiss' involvement with the visual arts, her use of poetic language in written and oral forms, her connections with the underground film. All of this is done while honoring the source and spark of weiss' whole career: poetry.

The volume: Academia and beyond

Though ruth weiss was active in a wide range of art forms – many of which are dealt with in the chapters in this volume – the poet always addressed the multimodal nature of her work as originating from her poetry. In *Full Circle* (2002), as a case in point, she plainly identified poetry as the source of all her artistic endeavors: "All of my work, the writing, performance, plays are rooted in poetry"[5]. Adhering to this premise, the present volume testifies to the centrality of the poetic vision throughout weiss' work. In this sense one can certainly ask, what better way could there be than to start a book about a poet through poetry?

Part I of this volume, entitled "Beyond Poetry", offers the reader an approximation to ruth weiss through poems, creative pieces, and speeches that act as homage to weiss at the same time that they provide insightful starting points to weiss' poetics. This section includes poems and tributes by poets such as

---

5 ruth weiss (2002). "'they call me a beatnik poet': ruth weiss Speaks to the Editor." *Full Circle / Ein Kreis vollendet sich*. Bilingual English/German. Transl. Christian Loidl. Vienna: Edition Exil. Here p. 166. Print.

Anne Waldman, Tate Swindell, John Wieners, Janet DeBar, Mary Norbert Körte or, among others, Jack Hirschman, some of them exclusively written for this volume. Seen through the eyes of these poets, weiss emerges in these tributes as a true poet and artist: a poet "since five / living bohemian love language of north beach" (S.A. Griffin), crafter of "lingual ecstasis" (Waldman), whose words are "volcanic magma solidified" (DeBar), and whose poems appear "darting like a mad flower's stamen" (Pete Winslow). Through the poetic portrayal of others, the reader gets cut-to-the-chase, candid depictions of weiss' artistic vision, such as the organic connection between her poetry, jazz, and bebop in particular, a connection that is analyzed in various chapters in the volume (see Chapters Six, Seven and Eight) and already advanced in this section by the poet A.D. Winans, who renders weiss' verse as "be bop jazz time / dances with timeless time". Conduits to weiss' essence as a poet, the poems in this section are also accompanied by one more text that allows the reader to access weiss' work from a less formal or academic path: the transcript of the speech delivered by Stefan Weber on the occasion of weiss receiving the Medal of Honor by the City of Vienna (2006), a touching tribute that weaves biography with attention to poetic detail. Kicking off the volume, all of these texts foreshadow the prevalence of poetry and the poetic vision in ruth weiss' oeuvre. They also frame weiss within an artistic and cultural sphere that is often diluted, if not directly ignored, in traditional scholarly studies.

Part II, entitled "Poetry, Jazz & Art", includes eleven essays and one interview written by scholars from varied backgrounds and nationalities. Grounding the analyses on weiss' multimodal work, these essays provide a comprehensive map of the extensive body of work produced by ruth weiss. The first two chapters deal with weiss' transnationalism and identity in movement. In "ruth weiss: Transnationalism and Resistance" (Chapter One), Benjamin J. Heal resituates and recovers the work of weiss in a transnational context of poetic experimentalism. Outlining the many facets of her life, art and writing, the chapter particularly focuses on her ongoing attack on the conventions of authorship and constructions of the singular literary genius through the use of contradiction, collaboration and various forms of multimedia expression. Heal's essay explores the resonances and more significant differences between weiss' work and the work of groups such as Problematica 63 in Spain, Noigandres in Brazil, Oulipo and Lettrism in France, PO-EX in Portugal, and Fluxus and L=A=N=G=U=A=G=E in the United States, rehistoricizing it within the wider framework of post-national experimental poetry.

In "'vienna. not quite.': Place, Movement, and Identity in ruth weiss' poetry" (Chapter Two), Stefanie Pointl considers some of the places and the routes between them in weiss' life to demonstrate how their poetic representation can be linked to her portrayal of trans-national connections. She argues that weiss' poetic movement between cities, countries, and continents questions the relevance of the concept of nationality for human coexistence – and the history of literature. In her poetry, weiss never explicitly engages with the issues of nationality and national identity, neither affirming nor dismissing them. In fact, her poems even rarely mention the names of specific countries. Instead, she questions their importance in more subtle ways. In this chapter, Pointl demonstrates how weiss rejects the notion of a self defined by nationality in favor of a postnational identity founded on experiences of movement and the resulting encounters rather than questions of origin. It constructs an identity that is shaped by multiple cultural and interpersonal connections but does not rely on identification with a particular nationality.

The next three chapters concentrate on poetry and weiss' poetic vision. Shifting from the notion of transnationality to gender, in "'how real is i?': Gender and Poetics in ruth weiss" (Chapter Three), Estíbaliz Encarnación-Pinedo utilizes the theoretical framework of gender studies and identity discourses to analyze poetry collections such as *Desert Journal* (1977) and *Light and Other Poems* (1976), as well as visual work such as *The Brink* (1961) to outline a consistent gender dissolution within weiss' work that is analyzed at the level of content as well as style. As such, the chapter considers weiss' development of fluid gender identities in poetry and art that is reminiscent of other poets associated with the Beat Generation such as Anne Waldman.

In "ruth weiss and the Poetics of the Desert" (Chapter Four), Polina Mackay focuses on *Desert Journal* (1977), which for many is considered the poet's masterpiece, to analyze three significant layers of weiss' book – namely the Judaic and Christian narratives, the desert as a poetic symbol and the desert as embodying America – to define and explore the poet's poetics of the desert. In this chapter, Mackay studies how all three layers complement each other to provide *Desert Journal* with an elaborate narrative of the desert as a site of spiritual searching, poetic discovery and a journey into American history, at the same time that it captures the creativity, the need for searching, the inconsistency, the poetics, ambiguity as well as the limits and the simultaneous potential of the mind.

Concluding this section on weiss' poetry, Chad Weidner examines the ways in which the work of ruth weiss connects to environmental discourse in "Reaching Towards the Light: Transitory Spaces and the Negated Material Body in Selected Texts by ruth weiss" (Chapter Five). With few exceptions, the environmental humanities have neglected the work of Beat writers entirely. This chapter uses weiss' example to illustrate how more transnational work is crucial to the continued development of both ecocriticism and Beat studies. Utilizing the application of contemporary ecocritical concepts, it challenges both the typical subjects of ecocritical research and common assumptions about the apparent hegemonic composition of the Beat Generation. More specifically, it examines extracts from ruth weiss' *South Pacific* (1959) and *Blue in Green* (1960) – two understudied collections – to divulge the many ways the texts engage in environmental discourse.

Chapters Six, Seven and Eight focus on the connection between weiss' poetry, the genre Jazz and Poetry, and performance. Hannes Höfer delves right into weiss' Jazz and Poetry in "Traditionally New: The Jazz and Poetry Work of ruth weiss" (Chapter Six). In this chapter, Höfer focuses on weiss' live performances of her poetry together with jazz musicians in order to highlight her role as an inventor, contributor, and inspiration of the genre. Concentrating on three different recordings which span six decades, Höfer analyzes the versatile and systematic relations established between the spoken word and the improvised music.

Much like Höfer's piece, Peggy Paccini's "Being tested: ruth weiss at the Summer of Love 2007" (Chapter Seven) closely explores the performance of poetry itself. In this chapter, Paccini investigates weiss' oral poetry by analyzing aspects of improvisation in her performances. Offering a close analysis of one particular performance of weiss, this chapter considers the usage of intonation, rhythm, and other elements of weiss' prosody in her performance at the 40th Anniversary of the Summer of Love event at Golden Gate Park, San Francisco, in comparison to other recordings from different periods and compares them in order to understand the poet's development from a bop-inspired recitation to a more open form that resembles the progression from Bebop to Free Jazz in music.

Chapter Eight presents an interview with ruth weiss conducted by Caroline Crawford. In this conversation Crawford documents weiss' early years in Berlin and Vienna, her escape from Nazism, her years at the Art Circle in Chicago, her life and work as poet, and her fusion of Jazz and Poetry. The chapter offers the reader firsthand observations not only into weiss' biographical history, but also with respect to the development of her poetic self. It provides readers with remarkable insights of how the poet's background flowed into her development of becoming a performing poet that featured a unique artistic approach and style.

Chapters Nine, Ten, and Eleven deal with weiss' expansion of poetry into other art forms. Frida Forsgren focuses on weiss' paintings in "ruth weiss and Visual Art: The Watercolor Haiku *A Fool's Journey* and *Banzai!*" (Chapter Nine). In this chapter, Forsgren studies the connection between weiss' work with the art scene in San Francisco in the late 1950s and early 1960s. Situating weiss as a prolific and eclectic poet and artist, the chapter analyzes two different expansions of poetry into watercolor painting to show how weiss' visual work displays affinities with the style, themes and energy of Californian Beat art.

Continuing the study of weiss' visual work, in "'Go to the roundhouse, he can't corner you there'" (Chapter Ten), Lars Movin focuses on *The Brink* (1961), a poetic film written and directed by weiss. In addition to analyzing the film – by focusing on the poem the film is mainly based on as well as the film itself – Movin also positions *The Brink* in the context of the art/film scene of the time, both within and outside the Beat Generation, as the film premiered in the wake of early underground works such as Robert Frank and Alfred Leslie's *Pull My Daisy* (1959) and Ron Rice's *The Flower Thief* (1960).

Steve Seid's "Luminosity Procured: A Photo-Essay" (Chapter Eleven) centers on the poet's participation in Steven Arnold's films. In particular, it focuses on *Luminous Procuress* (1971), a film in which weiss plays a major role and where she embodies an assortment of archetypal characters. The chapter includes previously unpublished slides by photographer Ingeborg Gerdes and an accompanying text about the poet's involvement in the film whose preserved version had its re-premiere in late April 2017 at the UC Berkeley's Pacific Film Archive. Seid explores how weiss functioned as both a muse and an active participant in San Francisco's artistic counterculture over the years.

Lastly, Thomas Antonic's "The ruth weiss Papers" (Chapter Twelve) offers a survey that provides an initial overview and detailed information on the scope of ruth weiss' literary estate. Offering a unique opportunity to delve into the archival material at the Bancroft Library at the University of Berkeley in California and weiss' former private archive at her home in Albion, this chapter provides scholars with information about the content and extent of published and unpublished written work as well as other documents such as audiovisual materials and photographs. It is intended, for future analyses, to make those who are performing scholarship aware of the vast amount of works ruth weiss created over the past seven decades that extend well beyond the scope of her published poetry collections which have served as the only real subject of studies to date. The chapter also briefly discusses some of weiss' theater plays.

Antonic's chapter is followed by a complete bibliography on ruth weiss compounded by him and Paul Pechmann. Divided in two sections – works by ruth weiss and literature about the artist – this bibliography rounds off the volume and offers a much-needed bibliographical source that helps to comprehend weiss' extensive career and output. The bibliography also includes a selection of relevant secondary sources on weiss, such as academic analyses and studies, interviews, portraits, and other documents.

Conceived as much as an academic reference book as a tribute and homage, this volume advances scholarship on one of the most active "Beat women" and post-World War II Bay Area poets while simultaneously proving to be a unique gem; more than a collection of academic essays, *ruth weiss: Beat Poetry, Jazz and Art* – including poems, interviews, transcribed speeches, and previously unpublished photos among others – serves as a defining example of weiss' excellence, a tribute to one of the most interesting figures of the post-war American counterculture. This volume is a testimony and tribute to the genius of a significant and unique artist. And this is just the beginning.

Acknowledgments

The editors would like to thank the contributors of this volume: poets who provided us with masterful and touching works of their art to pay tribute to their friend and colleague ruth weiss, as well as scholars who delved into their subject and spent a lot of time and effort to supply the world with a deeper knowledge of weiss' work and made this volume possible. We thank A. Robert Lee who contributes a preface for this volume and initially encouraged us to realize this book project. We thank Ann Cotten, Vishnu Dass (Steven Arnold Archive), Raymond Foye (John Wieners Estate), Nancy M. Grace, Owen Gump (Ingeborg Gerdes Estate), Susan McElrath (Bancroft Library), Jane Winslow, who provided us with information and material that was incorporated in this book, Sarah Earheart, Douglas Field and Juan Antonio Suárez Sánchez for their time and valuable feedback, and Mark Kanak who assisted the authors with the editing of the finished manuscript.

Last but not least, we thank ruth weiss who strongly supported this project from the very beginning by inviting us to work with her assistance in her private archive, providing us with an enormous amount of material such as unreleased audio and video recordings, manuscripts and hard to find publications, welcoming some of the contributors at her home, to give interviews, who granted us permission to digitize and reproduce her artworks and some manuscripts that were previously unpublished, et cetera. We are saddened that the poet was not able to see the fruits of our work due to her passing.

Estíbaliz Encarnación-Pinedo
Thomas Antonic

Part One: **Beyond Poetry**

Stefan Weber
Tribute to ruth weiss upon receiving the medal of honor by the City of Vienna in 2006

Hello everyone.

Dear ruth,
In your great book, *Desert Journal*, you write in the prologue:

> you are entering a certain desert
>
> like stones or bones
> marking sand
>
> flame & cloud
> things with wings
> call your number
> read your day
> see if it talks to you alone
> like stone or bone
> in sand
>
> other days
> other ways—

You are entering a certain desert:
 Entering a certain desert. They say that in the desert you can hear the blood rushing in your veins, that you're thrown back upon your own ego, so to speak, acoustically. And it's also said that there's silence in the desert. Silence isn't silence, isn't quiet, isn't noiselessness, isn't the opposite of something. Silence is absolute and sometimes difficult to bear, because silence is something incredibly intimate, like the sound of rushing blood.

 How quiet was it in your heart when you fled with your parents from Berlin to Vienna in 1933, fleeing the loud National Socialist noise in Germany? And later, in 1938, when drunken, intoxicated Austria loudly banned you from your new home as well? I can hardly imagine and could never really know.

 Fleeing is something you can't understand within yourself if you don't have to experience it yourself; it's just a hunch that involves something incredibly dangerous, oppressive and meaningless. Just powerlessness. Something inexplicable, something that excludes language.

But you, dear ruth, you have always found language, above all the language that came and comes from your silence. You were – as your translator Horst Spandler writes in an essay about you – "always there". Had arrived everywhere when others weren't there yet, so to speak.

> read your day
> see if it talks to you alone
> like stone or bone
> in sand

You've found the path to grandiose language, perhaps "fled" into language, but not aimlessly. Your goal has always been language and its musicality itself. Not finding words for something, but meaning.

At a time when public life, politics and many areas within culture are fleeing from the meaning of language, when political correctness eludes moral responsibility, when security is equated with fear, when "trust" and "respect" are abused, in short: when words lose their meaning – you are always in the middle and observe laconically:

> other days
> other ways—

And three dashes after ...

Where general flight from language leaves a great void, there you are, filling it up again.

One might even say that you're the happy Sysiphus who, once the stone has rolled down again, runs down the mountainside, liberated, only to push it up again. Liberated because the pushing up is always a new pushing up and thus makes a new kind of sense.

You are the center of the desert that seems to be turning around you, but sometimes something fleeting moves past you – things with wings – a dirty bird – maybe, and you pick it up and give the "thing" a meaning. And along with it, the desert, as well. In short: around you, there is no senselessness.

You write "Call your number": Call yourself – talk to yourself and ask yourself how you are. We should all do this, again and again, so that we can once again find a harmony with ourselves and return to where we come from: "Back to the roots". A return is only possible if you're one with yourself.

You've gone back, back to Berlin and you've come back to Vienna, too, visiting and cautiously, with your expansive, rich life in between in California, and you're astonished. In your company or while reading your texts one enjoys becoming a child again, a child with all his vigilance and curiosity and freedom of values. In your work, one discovers a new, free and mystical world, just emerging from this said silence. Where others confuse stillness with silence, it is your lightness of being, combined with this unbelievable seriousness towards this great adventure of life, that makes the reader so happy. You can't get enough awards and medals for that, and it's not at all possible to distinguish yourself in the actual sense, because it's *you* who distinguishes the world itself, Mother Earth.

Nevertheless, it is happening today – finally – through this formal act and that is good because actually, there are hardly words for it all.

I thank you and congratulate you with all my heart.

(Translation from the German by Mark Kanak)

Anne Waldman
RUTH IN THE HOUSE

"Everyone on this planet should have a share in the Universe" – Sun Ra

for ruth weiss

what rebellion, what *oddkin*s we are
witness to the fire, alterity, fantasy, dream density
collective jazz consciousness, arcane modus,
Anarchist poetry, foraging undercommons
Copernican heliocentricity?
Omniverse for poets! will luminous centers hold?
how to link up word and heart and panoramic attention
all the pores of listening in the house, all ears
ruth takes the microphone, eases our shy resistance
strikes the move, a pose, verbal acumen
an inner dance, chimeras of appetite and groan
all her ancient ornaments of grit and bone
a 20th century's struggle, a woman's struggle, thick and thin
born into war, rubble, extermination
born in crucible of time's mysterious test what is dished out
& 21st century's collapse in terror fear & dread the fate for
those "coming after" but she will keep tapping, riffing
witness ruth, the goddess the indomitable one
what lingual ectasis what trance and rhythm
what free radical haunts the premises and continues
in its wild sport and art but hers?

Tate Swindell
Future A.D. with Tate

after brief internal debate within as to whether typewriter or laptop keyboard the latter is flipped open and hopefully the words are still bouncing off of piano string vibrations of McCoy Tyner while Coltrane weaves in and out of traffic as Elvin Jones stethoscopic sounds pound pound pound & glance up near the stars bounds Jimmy Garrison fingertips can you see their pulse............................. there they are the notes of sheets of sounds that ruth weiss plucks from tho after reading poem to danny nicoletta written summer of 2017 for danny's first book of fotos—ruth took a poet's deeper breath—flashed far away eyes and exhaled gratitude to the muse who still visits—and I say yes! yes! yes!—the muse still visits—about every 6 weeks these days—hal is the co-pilot—tho sometimes ruth lets him get behind the wheel—dig this—just the other day—always Tuesday—when they arrive—same greeting where hal will offer a hug but don't touch his beer...yet—so hal is putting the gear away among which he stuffs the food in the fridge which meant chocolate chip cookies as well—well ruth had never heard of such a thing—and made it known—hal claimed to be full of surprises—ruth indulges while adding that when predictability enters he's out the door—quick draw hal says he might just hit the road right now—machetes at minimum to hack yr way outta this one—a few more pauses—call em beats if you want—hal sideways looks over—that grin pops up and he begins to laugh that laugh—and ruth— holding court—looking at him like she's still the one with the sharpest sword —and I tell hal that's a good one cuz they had me for a minute or half a minute or maybe it was just a fistful of heartbeats—hal bursting with laughs—he brings his own stash—and ruth says how much hal cracks her up—even when she doesn't want to laugh—this is when hal sticks his tongue out—hours to minutes later thinking about stubbing out this joint & the ashtray flash to hal as that's the only time joints are smoked and nostalgia drops in tho now it's just three weeks away and they'll be here—shit as a matter of fact exactly 3 and that same event will occur tho hal will be there—here/there all the same—fresh from el ay where i'm set to present 30 minutes on ruth—following George Herms—yet another circle back to ruth—footage cut from The Brink—readings with various band members—and kitchen shots shootin tha shit—her memory like a bank vault that is open to all—answers all questions which lead to more which lead to.........................sounds of Topanga Canyon reading with Benfaral Matthews......courtesy find of Thomas A. whom ruth always mentions as if he were a relative only separated by an ocean—tho frequent phone calls build bridges and reservoirs—another phenomenal connection orchestrated by the maestro—

https://doi.org/10.1515/9783110694550-005

ruth speaks about Benfaral, knowing he is still here—has seen him walking along the highway—felt the moment of his death of breath—the possibilities are limitless—time only happens to time—we know when to call—hearts and ears begin ringing while numbers dial in—5 nights ago—on the swings with Monika—I says—I gotta call ruth—tho we were puffing on a joint & baby it was San Francisco cold outside—the wind up in bernal heights has eyes & is a pervert—the joint half down—pull phone out of pocket—2 minutes ago!—as I was sayin it the phone was ringing—tho not ringing as mine is always silent—so I dialed back & she had info—further details into what was discussed days prior—fact checking & still gathering—and I also with questions & answers—had to share that I was at the Wild Side West—a relocated North Beach bar where ruth had bartended about 60 years ago—60 fuckin years mate! holy hell!—I rarely get to that bar tho did one time with ruth and hal when were all staying at the place I was watching for poet neeli cherkovski whom ruth & hal know & who happens to live a dozen houses away from the Wild Side—shit there's a line for neeli right there—oh yeah that's what I was saying—shit now I feel like I'm sitting in the kitchen with hal just riffin as he'll allow me the conch every now & then—well back to the wild side of ruth trading stories & then I tell her about lunch earlier that day with A.D. Winans—who turned 82 last month—and how he wanted to get together with the 3 of us not this time but the next time they come around—which as I type this—would be about early May—anyway ruth says—now wait a minute—hold on—I've got to write this down—always in that way—if you've ever heard it you can still hear it now—and there it is—across the line—floating thru clouds—and shit now that I think about it—that night dropped first rain in a long time and that's how ruth sent the line so quickly—*future A.D. with Tate*—holy shit! that's tops! what a gift! can't wait to tell her—in person—as there's nothing like seeing those green eyes flash beneath a frame of green hair and green fingernails and possibly a shimmering silver off the shoulder top—or black—usually black—tho stylish in a way unique to ruth—relaxed—curious with the scent of mischief lingering—that Cheshire cat can swing! you can listen to the record but ya gotta gotta—gotta see the way she commands the stage—the band watches one hand while she envelopes the audience with more—do not be fooled that there are only two arms at play—and at least a thousand eyes—thinking earlier of soup around the corner—mandarin style chinese—ruth's favorite—and I made reference to her wonton ladies line in poem recited hours before—and the smile that emerged from behind that bowl made me think of sphinx & pyramids & how ruth must have been there & yes! yes! yes! it all spirals around again—and now—writing this—I think of Harold—who was 92 when he floated on—and when I mentioned a birthday celebration this year—90—ruth at first opposed wanting to keep the momentum & natural flow—understandably so—tho later on there were words of perhaps a gathering—tho I see no reason why ruth cannot continue past 100...so I guess we can consider this to be continued...go light a green candle and write poetry or make a film or act in a film or paint or dance or share intimate conversation or just relax under the stars and listen to the trees breathe..........................well, just in case you have never attempted this before—find a tree—wrap your arms around it—press your left ear against it so that your heart also embraces—and

BREEEEEEEEEEEEEEEEEEEATHE

DEEEEEEEEEEEEEEEEEEEEEEEEEEEEEEEEEEEEEEEEEEEEEEEEEEEP

REPEAT

John Wieners
ARCHITECTURE

for ruth weiss

up north San Francisco/entrance exit
valley Bus Oakland Bridge, foothills Lax
rebout bay

never been as sailed
 Powell cable

Polk curry/ing prowurst
Rainy cancellation den snoozing

Janet DeBar
ruth

her complexity defies compression
she who is compression
power of compression
nouns and verbs triumphant
to hell with adjectives

her words look easy
they are hard
volcanic magma solidified
slick you risk slipping over them
sharp crystals cut you
as you try to get back up

cut to the core
cut to the chase
nothing to waste
every word counts
she counts every word
she counts

                                        June 8, 2014, the day after Mary Körte's 80th birthday party

Mary Norbert Körte
THE BIG NINE-OH

for ruth weiss

lady you startle fledgling ravens
out of the nest
they come flapping cawing
out of the fog
you feed them mightily daily
out of capacious pockets
crammed with seeds

                                      June 2018

Agneta Falk
RUTHFUL

for ruth weiss

there are three words in the English language
that describe you, ruth, perfectly:
RUTHFUL, TRUTHFUL & YOUTHFUL

the rhythm of that pulse of life
your turquoise head bopping
filling any room you enter
with a *je ne se quoi*

and you know something's alive
because the room's swinging
and your deep voice's timbre
rises from the floor, a steady
beat, beating through all those
false notes

ever TRUTHFUL

and that's sure a bendy road
you traveled along, with I'm certain
a lot of ducking and weaving too:
last train out of Austria
maybe the last boat to America
but so many *firsts* in your words
such spring in your feet

ever YOUTHFUL

and here you are, almost
a century young, with all
the grace of someone who's
made light of darkness
and shared it magically
around, without stopping
and still do, not missing
a beat

ever RUTHFUL

DU BIST IMMER SCHÖN UND WEISE

DU BIST ruth weiss!

S.A. Griffin
can't stop the beat

for ruth weiss

stream of consciousness
is how these things happen

berlin beat poet since five
living bohemian love language of north beach
moving picture of a poet on the brink
breaking the rules before the word

influential dreamer in full bloom
last of her line nerve-breaker
coming home

you have to drink and dance a lot
laugh alive on an old typewriter
with green hair

play a lot with words
in one's own voice
firmly rooted
in its center

a black jazz fool's journey
with big redwoods deep inside
this jazz and poetry
this connection with the
unbelievable

these stories in brilliant short time

an idea that you have no idea
to improvise divine haiku
into morning

impossible music always about to start
a hummingbird that never
leaves the heart

06/24/20

Pete Winslow
Soft Memories

for ruth weiss

1
little
but made of wonderful stuff
bloom the poems
darting like a mad flower's stamen

2
your vegetable sense of humor
blooms at my jokes

3
when elephants are burning alive
you don't think in terms
of meat

4
multiply the birds
on green sliderules
sing the note they make
on telephone wires

5
let us hover
on the very tip of the water
swimming a new stroke
which requires wings

6
the end of the belt
are in love

from the unpublished *POEMS! POEMS!*
Courtesy of Jane Winslow.

A.D. Winans
Grand Old Diva

for ruth weiss

she grooves with time
day time, night time
be bop jazz time
dances with timeless time
all rhythm no rhyme
birds in flight flap their wings
copulate with the wind

a magician's illusion where
time and words move from celibate
to shameless orgy
feed off the flesh of the other
pause in rollercoaster freeze
stop motion

she sings her song
another night
another day
bitch slaps father time
kaufman, son of jazz
in her heart
micheline in her blood
jazz in the Fillmore
jazz on the Harlem roof-tops
full moon rising
with poems that dig into my bones
lubricates the gears of my mind
lost in a haze of motionless motion

(11–21–12)

Jack Hirschman
Buon' Compleanno, RUTH!!

Neeli Cherkovski
In the desert

for ruth weiss

in the desert
        I would stand
over the ruins, mourning, and
pick up up a stone lion, and
wander to the border
of our estate, later, in the garden
sunlight will disappear
into the lemon tree, I loved you
and you abandoned me, one last time
the ashes of your lips, one final push
the walls subdued by the passing
of time, let me alone now, go away,
do not open your door

        let me be alone
and may I see clearly, I do not look for
your love, why bother? in the same manner
I never bow before the rain
even as every drop must be accounted for,
sanguine sound, shimmering knife of water
where the gate is forever open

come look come rain, join
in walking these ruins, oh make for me
a journal of life
amid stone facades, here a tarantula,
scorpion under cold rock, lizard
aware, rattler poised

I studied Medieval light,
the injuries are many,
light from the tomb
where love was born
we mean love in the realistic sense
love as we accept a candle
from Ibn Ara
love for paper and for poesy
in its varied applications

build for me a journal
of desert words

Steven Arnold
Day 25 (Diary excerpt)

SHE DAY FULL OF ENERGY AND PLAY. SHOT TWO SCENES WITH RUTH WEISS, MY POET AND TRAGEDIAN. SHE IS PURE MOONLIGHT. A MAGIC TALISMAN-LIVING, POSITIVE, CHARMED. WORKING TOGETHER IN ALL OF MY EARLIER FILMS HAS BUILT UP BETWEEN UP A DEPTH OF COMMUNICATION AND UNDERSTANDING, ALLOWING US TO WORK AT TIMES AS ONE. SHE PERCEIVES AND REFLECTS MY WORK PERHAPS MORE COMPLETELY THAN ANYONE ELSE. HER EYES TELL YOU SHE KNOWS ALL. PART EDITH PIAF, PART GIULIETTA MASINA, TRULY ONE OF THE MOST IMPORTANT WOMEN ON THE PLANET. I WOULDN'T MAKE A FILM WITHOUT HER.

Courtesy of Vishnu Dass (Steven Arnold Archive)

Part Two: **Poetry, Jazz & Art**

Benjamin J. Heal
ruth weiss: Transnationalism and Resistance

"POETRY IS A STRONG WEAPON." (weiss 2012)

Adorno's comment that it is "barbaric to write poetry after Auschwitz," (34) is particularly apt when applied to ruth weiss. Born in 1928 in Berlin to Austrian-Jewish parents, they escaped the Nazis in 1938 by moving from their home in Vienna to the United States, where she remained. Her recorded poetry's dissonant, contrapuntal relationship to the improvised jazz accompaniment marks a barbarism with word and sound that sets her work as a site of resistance to a simple construction of aesthetics in the Beat context. Her name, which she resolutely keeps all lower case, is another defiance in a career of defiance. As she states in a recent interview:

> Every time I sign my name, it is a revolutionary act, my way of standing up to the control of the "law and order" Germans in the '30's whose demand for control led to WWII and Nazis murdering millions of people, including my family. My name is a form of resistance. (weiss 2017)

This defiance was initially and surprisingly nurtured by mainstream cinema rather than underground poetry and Beat happenings. After the war weiss saw the Joseph Losey movie *The Boy with Green Hair* (1948), starring the 12-year-old Dean Stockwell, who would go on to star in movies such as Wim Wenders' *Paris Texas* (1984), and David Lynch's *Dune* (1984) and *Blue Velvet* (1986). Essentially an anti-war film highlighting the negative effect of war on children, after watching weiss decided to dye her hair green in solidarity with Stockwell's character, as a symbol of peace. In a twist of irony, Losey would become blacklisted and was forced to make his later films in Europe. These global interconnectivities and contexts are central to the approach to American literature advocated by Transnational scholar Paul Giles, who argues that American Studies today cannot ignore global contexts by "promulgating the virtues of particular kinds of identity politics as they emerge from the confines of the nation-state." (11) Instead he argues there must be a comparison between an American 'framework' and what lies outside it, thereby highlighting the specific elements that "reconstitute transnational networks in different ways." (11) This chapter will therefore examine weiss' life and work by disregarding specifically American contexts and will consider weiss in a more postnational context. This is particularly important at the current moment with the critical refiguring of Beat studies in a similarly trans- and post-national context, breaking the ingrained preoccupation within the academic study of Beat writers as a form of assimilatory American nationalism, from Jack Kerouac's illusory drive for the 'Great American Novel' to the Hollywood canonization of Allen Ginsberg's apparently intentional heroics in the history of American literary freedom. This status is in the process of being challenged with a more global focus of Beat studies beginning to take hold and challenging old paradigms, from Jimmy Fazzino's *World Beats* (2016) to the *Routledge Handbook of International Beat Literature* (2018) and Erik Mortenson's *Translating the Counterculture* (2018).

Even among the male Beat outsiders, and the overlooked female representatives, it is impossible to overlook weiss' uniqueness and marginality. And though she once considered her life, poetry and other work part of a forgotten history of Beat women, this has to some extent been rectified in recent years, with increasing recognition of her work and that of the other female Beat writers in collections such as Brenda Knight's *Women of the Beat Generation* (1996), Mary Paniccia Carden's *Women Writers of the Beat Era* (2018) and the edited collection *Out of the Shadows: Beat Women Are Not Beaten Women* (2015).[1] Although collections such as *Kerouac on Record* (2018) compound the failure to acknowledge her role in the development of Beat poetry and music crossovers, in its own right her work is gaining increasing recognition as weiss herself noted in a 2012 interview:

> I made a film in 1961 [*The Brink*], the only one I directed actually. I've been in many movies, but that's the only one I directed and it's one of my poems; it was very little paid attention to, but all of a sudden it's now ... it was in Venice, at the Whitney Museum in New York, last week it was at the Hudson Art Fair, somebody showed it in Ghent ... all of a sudden. (Antonic)

Because of Kerouac's standing in the canon of Beat literature, weiss' friendship with him, which began before the success of *On the Road* (1957), has overshadowed her own work. Moreover, her role in nurturing Kerouac's nascent interest in the haiku form has been overlooked, most notably in *Book of Haikus* (2003). As Knight notes, when weiss and Kerouac would meet at the Wentley Hotel, Kerouac would comment "you write better Haiku than I do." (1996, 246) Kerouac was, unfortunately and unlike Ginsberg, not a great champion of his friends' work, and

---

[1] Other collections include Richard Peabody's *A Different Beat: Writing by Women of the Beat Generation* (1997) and Nancy M. Grace and Ronna C. Johnson's *Breaking the Rule of Cool: Interviewing and Reading Women Beat Writers* (2004), and their edited volume *Girls Who Wore Black: Women Writing the Beat Generation* (2002).

weiss continued to give readings and publish limited run editions of her poetry without the success or recognition of her friend. As Jimmy Fazzino notes in relation to the canonized Beats:

> The Beats were prolific travelers, and the fact that they produced some of their most significant and enduring works abroad [...] suggests that their calling as writers was somehow predicated upon their leaving the United States behind. This distance from home is what opens up a space for all sorts of unexpected connections and crossings to arise in their work. (2)

The irony for weiss is that she was already away from home when she arrived in the United States, and this gives her work an intellectual advantage over the other Beats; it is always, already, grounded in rootlessness.[2]

weiss and Women

American women of the 1950s are often presented as being confined to the boxes assigned by a repressive society. Knight for instance notes that "Women of the fifties in particular were supposed to conform like Jell-O to a mold. There was only one option: to be a housewife and mother [...] being Beat was far more attractive than staying chained to a brand new kitchen appliance." (1996, 3) Yet this is oversimplifying matters, and one can see in weiss' case a difference and uniqueness of upbringing and attitude, a truly transnational identity and poetics that is both in harmony with and a rejection of Beat poetics. Indeed the androcentrism and silencing of women that occurs in *On the Road* is turned on its head by weiss. It would have been quite simple to produce work referencing her time with Beat luminaries Kerouac and Neal Cassady, but instead she publishes *Gallery of Women* (1959), a sublime meditation on and celebration of the women artists in her immediate sphere. *Gallery of Women* is not merely resistant to the centrality of men in Beat poetics, but it also challenges the centrality of poetry itself with illustrations of the women, in a resistant nod to the pin-up centerfold, placed in the center of the book. The achievement of *Gallery of Women* is to re-center the feminine that was historically 'Othered', as noted famously by Simone de Beauvoir – "Thus humanity is male and man defines woman not in herself but as relative to him." (46). The poems are all portraits of women, and weiss also includes Sutter Marin's Surrealism inspired visual portraits of some of them in the middle of the book. Following Pierre Macherey's

---

[2] weiss states she feels at home in Vienna (2012).

dictum, that "what is important in a text is what it does not say" (85), it is evident that these poems of women effectively and deliberately seek to portray them without reference to men. There is even a lack of male or masculine imagery in the poems, which is a tribute to weiss' resolute skill. The enigmatic absences effectively point to the obvious oppression of women by men in American society. The only male imagery in the book, the rooster in "Cockie with Earth-Paintings", is resolutely sexualized, reversing the usual sexual objectification of women for male pleasure. The use of the word "gallery" in the title is also quite deliberately drawing attention to objectification. The center(-fold) portraits, drawn by a male artist, exemplify this aspect of the book, with the images of women de-sexualised into deconstructed, complex surrealist forms, the male gaze of the artist is therefore denied simple pin-up objectification. In the same way the poetry reconstructs the notion of a gallery, from women as the formal nudes of classical art, to the more complex representation of women through words. The rest of the book largely focuses on female and feminine images of nature, birds and flowers, with the beautiful image of the naked woman riding a sea horse (a species where the male gives birth to the young) on the front cover. This image was produced by weiss' husband Mel Weitsman, which marks a turn in her work towards a more transgressive and androgynous approach to representations of gender and sexuality and complements her open sense of post-nationalism.

Todd F. Tietchen's androcentric essay on Beat transnationalism in *The Cambridge Companion to the Beats* unfortunately fails to mention weiss' clear centrality in terms of a transgressive, androgynous and transnational presence in the Beat canon. As his essay gets bogged down in cosmopolitanisms and contemporary oppositional politics in the acceptance of Beat literature in academia, the failure to examine weiss' female perspective on transnational poetics looms large (221–222). In transnational terms weiss finds herself in a curious position, joining the United States essentially as a refugee at the beginning of its journey to world dominance she found herself in a dichotomous position, feeling a sense of rootlessness, living in a home of sorts and pleased with the acceptance, and yet also rejecting the perceived enforcement of counterintuitive ideologies through her poetic promotion of peace, acceptance and cohesion in the context of dialectical change. These tensions are beginning to appear in the resistances of *Gallery of Women* and become ever more apparent in the following collection, *Blue in Green*, which seeks to hone down word and image to the absolute minimum, and eschews rhyme while embracing abrupt changes and assonances. There is a direct reference to the need for generational cohesion which reads like Bob Dylan's "The Times They Are A-Changin'" (1964):

> we are of man and woman
> and having named us
> both will still
> use
> us as a battlefield (weiss 1960)

The reference to "man and woman" shifts to reference "us" suggesting a post-gender sense of cohesion, with generational tensions and literal warfare being the central problems to overcome. The rhythm and musicality are also more apparent, with phrases such as "warm swarms the wind / and wide / open-arms /a-voice", which highlights a developing sensibility that would continue with her interaction with jazz.

weiss and Jazz

weiss' embrace of jazz can be seen as a further form of resistance to both mainstream forms of music and the formal structures of traditional poetics, recalling the Harlem Renaissance poetry of Langston Hughes, whose syncopated rhythms resonate with weiss', particularly this section of "Weary Blues" (1929):

> He did a lazy sway... .
> He did a lazy sway... .
> To the tune o' those Weary Blues.
> With his ebony hands on each ivory key
> He made that poor piano moan with melody.
> O Blues! (50)

Compare the rhythm here with weiss' "Sena" (1976):

> gotta go for a day-break walk
> gotta go for a day-break talk
> don' know where will walk about
> don' know what will talk about
> saw those leaves flutter
> saw the candle flicker
> no wind (10)

Her enthusiasm for the often-overlooked form of jazz poetry is unsurprising given the Beat interest in African-American culture. Her use of the form has similarities with the jazz inflected prose of the best sections of Kerouac's *On the Road*, yet for weiss there is a politics of individual resistance to conformity that remains unsatisfactorily unresolved in Kerouac's novel. For weiss there can be no compromise to the form, in that her poetics of resistance and commitment is not about to try to appeal to mainstream sensibilities. Indeed her interest in jazz and Black culture was not a passing fad, or case of cultural appropriation, it became fundamental to her poetic composition and the focus of her poetic performance. Her other Beat friend of the period, Bob Kaufman, took that commitment to great lengths too, with weiss acknowledging his influence in her 1986 poem "For Bobby Kaufman" – "crossed your bridge / with your big word / and your huge silence" (2011, 53).

Kaufman's work was often set to jazz rhythm such as "Bagel Shop Jazz" ("Talking of Bird and Diz and Miles" [15]), and his influences were, like weiss', Dada and Surrealism inflected. His sense of a post-racial world also echoes in weiss' politically and artistically committed poetry. Kaufman's commitment to poetry as performance, and the power of silences, is underlined in weiss' reference to his 10-year vow of silence following John F. Kennedy's assassination. And yet weiss' abstract images also feel disconnected from both politics and their historical contexts. Preston Whaley Jr. contextualizes weiss' status in *Blows Like a Horn: Beat Writing, Jazz, Style and Markets in the Transformation of U.S. Culture* (2004), where he writes: "More than most of the Beat writers, weiss' art marks the collision of white biography on the fringe with African American culture." (65) Eschewing the rose-tinted primitivism of Kerouac's representations of Black culture – the "happy negroes of America" that James Baldwin rejected as "absolute nonsense, and offensive nonsense at that: I would hate to be in Kerouac's shoes if he should ever be mad enough to read this aloud from the stage of Harlem's Apollo Theater" (231) – weiss felt entirely comfortable embedded in Black culture, as she explains:

> And there were so many black people I was connected with, you know, dancers and musicians and poets, and they happened to be black. The first black person that I was close to was in my grammar school here in Vienna. And my other connection is that my parents put me into this children's home in Harlem. And I was the only white girl in my class, and I couldn't even speak English. And so I've always been at home with their language and their music and most of their jazz musicians. (Antonic)[3]

This comfort is a profound rejection of the fascism that she escaped in Europe, and is free of the patronizing platitudes of *On the Road*. weiss' race politics have

---

[3] weiss also mentions this in her autobiographical poem "I Always Thought You Black" in *Can't Stop the Beat* (2011).

been experienced first-hand. This deep sense of authenticity is central to weiss' sensibility, and it comes through in her poetic practice.

weiss' Poetics

Nancy Grace categorizes weiss' poetics as "a spontaneous method of free association linked to Yeatsian automatism and Tzarian Dadaism," (59) pointing to the deeply embedded transnational range of influence on her work, but Grace confusingly points to a "Romantic and Buddhist belief in 'first thought is best thought'." (59) weiss was connected to Buddhism, but Buddhist belief and practice are not evident in her work per se, and such a statement gives the impression that weiss writes automatically without meditation, thought and reflection – like a Kerouac on Benzedrine – and this is patently not the case as she states in a recent interview: "My advice is 'less is more.' Never overwrite or just keep writing. Spare use of language, only using the right words will lead to better work, get you closer to greatness." (2017) As with jazz improvisation it is a skill practiced and learned which follows the structure of the song being played, which seems a better analogy of weiss' controlled spontaneity. This is born out by her performances, where she reads her poems in time with the music rather than performing entirely free improvised poetry or scatting. As weiss states, in such moments "the meaning and the reverberation of the sound have to hit at the same point" (qtd. in Grace, 59). This is also noted by David Amran, the jazz musician who regularly collaborated with Kerouac:

> We never once rehearsed. We did listen intently to one another. Jazz is all about listening and sharing. I never drowned out one word of whatever Jack was reading or making up on the spot. When I did my spontaneous scatting [...] he would play piano or bongos and he never drowned out or stepped on a word or interrupted a thought that I or anyone else had when they joined us in these late night-early morning get-togethers. We had mutual respect for one another, and anyone who joined us received the same respect. (2003)

Ironically, given all the attention I have paid to the interconnectivity between weiss' work and jazz, silence is central to her written work, where the large blank spaces on the page of her poem "Forty-One Dragon-Steps" contrast with the babbling lyricism of the preceding "One More Step West Is the Sea" in her debut collection *Steps* (1958). In music a step is the difference in pitch between two consecutive notes of a musical scale, so again weiss is setting her work as an in-between, a liminality, forming a contrapuntal relationship between points. The silence marks not only the silence of women's voices, and their marginality even within the counterculture Beat Generation circles of San Francisco of the late 1950s, but also a transience, the movement between spaces, and a form of Deleuzian deterritorialization marked at the end of "One More Step West Is the Sea" by the comment that "I've lived here five years / and only met two who were born here" (1958, 7). This is followed by an all-capitalized line of the title of the poem, powerfully underlining the limits of America, how they have perhaps been met by, or are about to be cast-off into the sea across the Pacific. The poem then ends on the next page with a powerfully silent ellipsis – its power ironically denoting the powerlessness of the individual in the face of that crushing manifest destiny. In contrast with "One More Step West Is the Sea", the "Forty-One Dragon-Steps" are Haiku-like impressionistic fragments, many just two or three lines per page. Their clearly Sinospheric influence from the dragon imagery to the meditative images they evoke point toward the 'Eastern' influences that impacted on the work of the other Beat writers. But again, weiss' collection is divided into two, the urban, city-based images of "One More Step West Is the Sea" juxtaposed by the Eastern imagery and Haiku-like form of "Forty-One Dragon Steps". They combine to point towards a post-national formulation embracing both. The use of the word "pearl" three times; "the pearl of becoming … […] the pearl / from on high … […] the pearl of becoming—" (1, 2, 41) in "Forty-One Dragon Steps" evokes the image of conflict with the attack on Pearl Harbor marking the key point in contemporaneous history when the West came into direct conflict with the East, an apparently evocative embrace of Hegelian dialectic in her work.

weiss' barbarism of form, through her use of dissonance and rejection of traditional poetic rhyme and meter, can be seen to have more in common with William Burroughs and Brion Gysin's later Cut-up method than Kerouac's Westernization of Haiku. Kerouac felt that Western languages could not adapt to the "fluid syllabic Japanese" form, proposing that "the 'Western Haiku' simply say a lot in three short lines in any Western language. Above all a Haiku must be very simple and free of all poetic trickery and make a little picture and yet be as airy and graceful as a Vivaldi Pastorella" (Kerouac 2004, x–xi). weiss certainly takes the Haiku away from its Japanese origin and makes it her own by imposing an almost Lettrist reduction of semantic value from the poem. In attacking the rhythmic flow of speech by employing a scattershot improvised jazz approach, weiss' poetry conforms to the more dissonant directions of the avant-garde progression of jazz as it crossed the Atlantic in the 1960s and 70s through the work of Peter Brötzmann and Albert Ayler. One example of this is "Cockie with Earth Paintings" from *Gallery of Women*, which also references this dissonance and structural violence with a coded reference to fellatio. Its sexually charged staccato rhythm and contrapuntal use of two columns (representing both partners) is both musical in an avant-garde jazz sense, and violent in that the use of columns acts like one of Burroughs' three column cut-up compositions:

```
the dark
she waits
the profile
is explode
my mouth
to fill
the question
i am to about
to loose the                if I were the
light                       rooster to
am dark                     wash my free hand
                                i would
                                to feather off
                                night
DAWN                        The open mouth
the rooster is shrieking
to
let
go
grain
```

While the staccato rhythm enhances the sexual reference, this central section of the poem demonstrates weiss' mastery of dissonance. The complete absence of assonance in this section is striking, as are the grammatical breaks – "the profile / is explode" – which give an effect of the text having been cut. The short lines emphasize the effect, while the offset section is also deliberately not justified to the rest of the text, making it unclear whether the reader should read across or down or insert the text in the gap. The trinity of possibilities marks a point of charged significance where weiss' poetics effectively and deliberately resist simple semantic explanation.

weiss and Haiku

That weiss and Kerouac became interested in Haiku at the same time is also fascinating. Given the level of anti-Japanese sentiment and propaganda in the United States during the war, the interest in the East demonstrated by Kerouac and Ginsberg, and more directly via Philip Whalen and Gary Snyder, may be a little surprising. There were four significant books of Haiku translations and commentary by Robert L. Blyth published between 1949 and 1952. These became the basis for the interest in Haiku of the Beat poets. From reading the list of the published works, it appears that no women poets were involved with Haiku at this time. Yet weiss and Kerouac wrote their Haiku together, often using it as dialogue between them, although none of this was published (Reichhold 1986). Haiku is a particularly relevant form for weiss, because, as Yoriko Yamada states, the essence of Haiku is cutting (*kiru*) (255). This is captured in poetic form by the juxtaposition of two images or ideas with a cutting word (*kireji*) between. This punctuates the juxtaposed elements, marking the moment of separation and shaping the interrelations. The cutting element is both a reflection of the barbarism of word that weiss' poetry produces, but also acts metaphorically in terms of the broader identity politics that are weiss' preoccupation. Gender, race, national identity, all marked by a triangularity of Others pushing towards a oneness reflected in weiss' poetry, painting, performance and film.

weiss and Collaboration

This transgressive and transnational context of weiss' transmedial poetic experimentalism balances out the many liminalities in her life, art and writing. Her sense of being in-between, the *kireji* of the Haiku, as a result of her trans-national identity with a particular focus on her ongoing attack on the conventions of authorship and constructions of the singular literary genius through the use of contradiction, collaboration and various forms of multimedia expression. Collaboration was also an important facet of Burroughs' work:

> BURROUGHS: A book called Think and Grow Rich.
> GYSIN: It says that when you put two minds together...
> BURROUGHS: ... there is always a third mind ...
> GYSIN : ... a third and superior mind ...
> BURROUGHS: ... as an unseen collaborator. (19)

This sense of collaboration, both with its Nazi undertone of complicity with fascism and its more positive aspect of countering contemporary constructions of the solitary genius in artistic construction, which also happens to be triangular, is particularly appropriate for weiss' longstanding commitment to collaboration. The notion of collaboration as a "third-mind" connects to another of Burroughs' preoccupations; language. Further transnational liminalities appear when considering weiss' use of language, as her first poem, written at age 5, was written in German:

My first poems in German, and then I came to America. For a while I couldn't speak English, I was just 11. I didn't write a lot. Then at the age of 12 I started writing a lot, all in English. And I like the English language so much, for writing, more than all the others. (2012)

weiss continues to publish in both English and German despite her preference for writing in English. This linguistic ambivalence or oscillation makes her work transgressive at the level of language, and shows her awareness of the unique qualities this space between languages enables. It is not clear to what extent weiss' poems are affected by her background with German, but in the same interview she mentions feeling at home in Vienna, despite making her home in America. The title of her film *The Brink*, is yet another reference to inbetween-spaces, much like the poem "One More Step West Is the Sea" in *Steps*, and the work in *Desert Journal* (1977). A collaboration with painter Paul Beattie that Stan Brakhage describes as "one of the most important San Francisco films of the period", (qtd. in Carney) *The Brink* begins with the image of a caterpillar, whose life-cycle is a trinity of caterpillar, chrysalis and butterfly. It continues with a couple walking in a street, cutting to a man in a coffee shop with his back to a mirror, wearing a woman's hat. This liminality between image and reflection, gender binaries and life cycles is foregrounded by weiss' poetic voice performance, which alludes to reflexivity in the poetic formation of city-spaces "city up, city down".

Often considered her most important work, *Desert Journal*, has that status because it encapsulates many of the themes of the earlier work. *Desert Journal* lends itself to a wide multi-media impetus ranging from her own performances to the Paul Blake drawings included in the collection. It is also clear that weiss' performances had, at least at one stage, a very strong visual component. The cover of *Can't Stop the Beat* (2011) presents weiss during one of her film collaborations with Steven Arnold in the late 1960s, with face painted white, painted on eyebrows and a skullcap. Her strikingly androgynous appearance recalls the dyeing of her hair green in the early 1950s as a protest. Her 'total' trans- or multimedia approach to poetry, through performance art, painting, and film resonates with European and American movements such as Dadaism, Surrealism, Lettrism, Situationism, *Problematica 63* in Spain, *Noigandres* in Brazil, *PO-EX* in Portugal, and *Fluxus* and $L=A=N=G=U=A=G=E$ in the United States. (Ledesma) Yet despite her embrace of collaboration she was never considered a member of any group or collective and has only recently adopted and accepted the Beat moniker.

Desert Journal again recalls weiss' obsession with liminality, the desert itself an uninhabitable geographical region between urban or verdant regions, a hostile place of discovery and becoming not a final destination. As Grace points out, "the presence of the titular 'forty' evokes such central Biblical stories as those of Jesus' forty days and nights in the desert and Moses' forty years seeking the promised land." (60) It also recalls Paul Bowles' "Baptism of Solitude" (1953), an experience of self-discovery that occurs in the Sahara, which he describes as:

> [...] a unique sensation, and it has nothing to do with loneliness, for loneliness presupposes memory. Here, in this wholly mineral landscape lighted by stars like flares, even memory disappears; nothing is left but your own breathing and the sound of your heart beating. A strange, and by no means pleasant, process of reintegration begins inside you, and you have the choice of fighting against it, and insisting on remaining the person you have always been, or letting it take its course. For no-one who has stayed in the Sahara for a while is quite the same as when he came. (75–76)

In a similarly enigmatic opening, *Desert Journal* begins:

> you are entering a certain desert
> like stones or bones
> marking sand
> flame & cloud
> things with wings
> call your number
> read your day
> see if it talks to you alone
> like stone or bone
> in sand

The metaphorical desert in *Desert Journal* is a space of renewal, where readers can choose a day and see if it speaks.[4] Deliberately un-paginated, the effect is of groundlessness. The poems act like Buddhist mantra, meditations of free association which nevertheless resonate with references. For instance the first day contains reference to Edgar Allen-Poe's "The Raven", along with the aforementioned androgyny of weiss' work and performance – "either sex / or both / what is the hermaphroditic fact?". Birds are a recurrent symbol throughout weiss' work, and they symbolize women in different forms, with the metaphor of flying underscoring the motif of freedom and resistance to any form of conformity in her work. This occurs in *Blue in Green* where the clear metaphorical associations of blue with the sky and the ocean, green with foliage and nature are contrasted with the blackbird:

---

4 The performance requests members of the audience to choose the day.

> And the black birds fly
> black black bird
> you who leave the others
> in one moment scoop
> and bring the dusking sky back
> to the rest in flight (1960)

The black bird here represents the Beat poet, breaking free of patriarchal expectations and revealing the truth (sky) in flight (poetry). The flight metaphor also represents travel and transnational connections – making poetic links to traditions from Europe, Japan and America.

weiss' largely post-national life and work can be seen as a resistance to fixed labels, conventions and ultimately, power. Her transmedia, or "total" approach to poetic expression extends this resistance. Yet this is arguably balanced by the avoidance of didacticism and polemic in her work, in contrast to the work of Beat poets such as Allen Ginsberg, through a more Modernist sense of vagueness and impressionism. Yet weiss' work is so finely carved with clear poetic focus, several of the book titles, such as *Gallery of Women*, point to a straightforward and open approach to meaning, while her preferred context, performance, provides a focal point that aids political understanding of her work through its many liminalities, transnational and trans-cultural influences and finally its quality of a profound resistance to fixed semantic readings, retaining her status as a Beat outsider.

Works Cited

Adorno, Theodor W. "Cultural Criticism and Society." *Prisms*. Cambridge, MA: MIT Press, 1983, pp. 17–34.
Amram, David. *David Amram: Poetry and All That Jazz,* 2003. Web. https://http://www.allaboutjazz.com/david-amram-poetry-and-all-that-jazz-david-amram-by-david-amram.php?width=1440. Accessed 21 Apr. 2018.
Antonic, Thomas. "Vienna Never Left My Heart." A conversation with ruth weiss. *European Beat Studies Network, 2012*. Web. https://ebsn.eu/scholarship/interviews/ruth-weiss-interviewed-by-thomas-antonic/. Accessed 21 Apr. 2018.
Baldwin, James. *Nobody Knows My Name: More Notes of a Native Son*. New York: Dell, 1975.
Beauvoir, Simone de, Constance Borde, and Sheila Malovany-Chevallier. *The Second Sex*. New ed., transl. by Constance Borde and Sheil Malovany-Chevallier. London: Jonathan Cape, 2009.
Bowles, Paul. *Travels: Collected Writings, 1950–93*. London: Sort Of, 2010.
Burroughs, William S., and Brion Gysin. *The Third Mind*. New York: Viking, 1978.
Carden, Mary Paniccia. *Women Writers of the Beat Era: Autobiography and Intertextuality*. Charlottesville: University of Virginia Press, 2018.

Carney, Ray. *The Beat Movement on Film a Comprehensive Screening List*. UC Berkeley Media Resources Center, 1996. Web. http://www.lib.berkeley.edu/MRC/CarneyFilms.html. Accessed 21 Apr. 2018.
Fazzino, Jimmy. *World Beats: Beat Generation Writing and the Worlding of U.S. Literature, Re-Mapping the Transnational*. Hanover, New Hampshire: Dartmouth College Press, 2016.
Forsgren, Frida, and Michael J. Prince, eds. *Out of the Shadows: Beat Women Are Not Beaten Women*, Kristiansand: Portal, 2015.
Giles, Paul. *Virtual Americas: Transnational Fictions and the Transatlantic Imaginary*. Durham, N.C. London: Duke UP, 2002.
Grace, Nancy M. (2004). "Ruth Weiss' *Desert Journal:* A Modern-Beat-Pomo Performance." *Reconstructing the Beats*, edited by Jennie Skerl, New York: Palgrave Macmillan, 2004, pp. 57–74.
Johnson, Ronna C., and Nancy M. Grace. *Girls Who Wore Black: Women Writing the Beat Generation*. NJ: Rutgers UP, 2002.
Hughes, Langston. *The Collected Poems of Langston Hughes*. New York: Vintage, 1994.
Kerouac, Jack. *On the Road*. 1957. London: Penguin, 2000.
Kerouac, Jack. *Book of Haikus*. Ed. Regina Weinreich. London: Enitharmon Press, 2003.
Knight, Brenda. *Women of the Beat Generation: The Writers, Artists and Muses at the Heart of a Revolution*. Berkeley, Calif.: Conari Press, 1996.
Losey, Joseph, dir. *The Boy with Green Hair:* RKO Radio Pictures, 1948.
Ledesma, Eduardo. *Radical Poetry: Aesthetics, Politics, Technology, and the Ibero-American Avant-Gardes, 1900–2015*. New York: SUNY Press, 2016.
Macherey, Pierre, and Geoffrey Wall. *A Theory of Literary Production*. London: Routledge and Kegan Paul, 1978.
Mortenson, Erik. *Translating the Counterculture: The Reception of the Beats in Turkey*. Carbondale: Southern Illinois UP, 2018.
Reichhold, Jane. "Those Women Writing Haiku." *AHAPoetry*, 1986. Web. https://ahapoetry.com/TWWHBK.HTM. Accessed 21 Apr. 2018.
Tietchen's, Todd F. "Ethnographies and Networks: On Beat Transnationalism." *The Cambridge Companion to the Beats*, edited by Steven Belletto, Cambridge: Cambridge UP, 2017, pp. 209–224.
Warner, Simon, and Jim Sampas, eds. *Kerouac on Record: A Literary Soundtrack*. London: Bloomsbury, 2018.
weiss, ruth. *Steps*. San Francisco: Ellis Press, 1958.
weiss, ruth. *Gallery of Women*. San Francisco: Adler Press, 1959.
weiss, ruth. *Blue in Green*. San Francisco: Adler Press, 1960.
weiss, ruth. *Light, and Other Poems*. San Francisco: Peace & Pieces Foundation, 1976.
weiss, ruth. *Desert Journal*. Boston: Good Gay Poets, 1977.
weiss, ruth. dir. *The Brink*. 1961. Albion, Ca, 2008.
weiss, ruth. *Can't Stop the Beat: The Life and Words of a Beat Poet*. Studio City, CA [Los Angeles]: Divine Arts, 2011.
weiss, ruth. "Conversation with Beat Poet ruth weiss About the Cellar, Blues, Jazz, Sappho, Beats, and Poetry." Interview by Michael Limnios. *Blues GR*, 2012. Web. http://blues.gr/profiles/blogs/conversation-with-ruth-weiss-about-the-cellar-blues-jazz-sappho. Accessed 21 Apr. 2018.

weiss, ruth. "Interview – ruth weiss." Interview by Brenda Knight. *Women's National Book Association, San Francisco Chapter*, 2017. Web. http://wnba-sfchapter.org/interview-ruth-weiss/. Accessed 21 Apr. 2018.

Whaley, Preston. *Blows Like a Horn: Beat Writing, Jazz, Style, and Markets in the Transformation of U.S. Culture*. Cambridge, Mass.; London: Harvard UP, 2004.

Yamada, Yorik. *Haiku East and West: A Semiogenetic Approach*. Bochum: Brockmeyer, 1985.

Stefanie Pointl
"vienna. not quite.": Place, Movement, and Identity in ruth weiss' Poetry

"This is not a travelogue," (25) ruth weiss writes in *Can't Stop the Beat: The Life and Words of a Beat Poet* (2011), yet her autobiographical poems tell of a life lived in seemingly constant motion. Escaping the Nazis twice, first from Berlin to Vienna in 1933 and then narrowly to New York in 1938, travelling solo around post-war Europe, hitching rides across the American continent and eventually settling in San Francisco—these are only some of the routes on which weiss' life has taken her. Readers of her poetry will inevitably be struck by her understated yet profound engagement with her past and an enduring fondness for Vienna. It might seem surprising to find such a lasting connection to the Austrian capital in the work of a Nazi refugee. However, far from dwelling on a bygone era, her poems contain neither sentimental nostalgia for a lost home in Europe nor unbridled enthusiasm for the adopted one in the United States. Instead, they convey a voice that moves freely within and between all these places and does not quite allow itself to be tied to any of them.

This chapter considers some of these places and the routes between them to show how their poetic representation can be linked to weiss' portrayal of transnational connections. I argue that weiss' movement between cities, countries, and continents questions the relevance of the concept of nationality for human coexistence. In her poetry, weiss never explicitly engages with issues of nationality and national identity, neither affirming nor dismissing them. In fact, her poems rarely mention specific countries. Instead, she questions their importance in more subtle ways. weiss' poetry rejects the notion of a self defined by nationality in favor of a postnational identity founded on experiences of movement and the resulting encounters rather than questions of origin. It constructs an identity that is shaped by multiple cultural and interpersonal connections but does not rely on identification with a particular nationality.

Transnationalism and National Identity

In this age of global interconnectedness and unprecedented migration,[1] transcultural connections are gaining increasing significance as an area of research. Tra-

---
[1] According to the United Nations *International Migration Report 2017*, the number of interna-

ditional notions of cultures as spatially and temporally fixed sets of practices, customs, and beliefs shared by a group of people in a specific geographical location (Welsch, 194) are becoming superseded by an understanding of "culture as a fluid, mobile, transnational phenomenon that predates and often ignores nation-state boundaries" (Jay, 57). With cultural practices dissolved from national contexts, the concept of nationality no longer necessarily serves as a source of identification for people with ties to multiple locations due to migratory movement. In 2002, Paul Giles posited that "American studies today cannot continue to operate in the same old way, simply promulgating the virtues of particular kinds of identity politics as they emerge from the confines of the nation-state" (11). This statement is part of the so-called transnational turn in American studies, heralded by Shelley Fisher Fishkin's presidential address to the American Studies Association in 2004 ("Crossroads of Culture") and exemplified by publications such as *Globalizing American Studies* (2010), *Re-Framing the Transnational Turn in American Studies* (2011), and *The Cambridge Companion to Transnational American Literature* (2017). According to Fisher Fishkin, research on transnationalism, among other aspects, aims at investigating the role of international movements in establishing connections beyond national borders (24), as well as highlighting "the broad array of cultural crossroads shaping the work of border-crossing authors" (32).

In recent years, this awareness of cultural interconnectedness and the importance of transnational movements has also called into question the study of literature within national boundaries. The pluralistic reality of contemporary societies stands in sharp contrast to traditional literary canons and their claim to reflect a nation's unified identity founded on shared ideological values and aesthetic ideals (Jay, 33). In light of the decisive role of migration and cultural hybridity, especially in the history of American literature, there have been calls for alternative approaches to literary studies, for example by Paul Jay who suggests considering literature within linguistic rather than national frameworks. He views Anglophone literature as "increasingly postnational," considering that English is used as the language of creative expression by artists around the world and from a variety of cultural backgrounds (Jay, 33). Apart from acknowledging these present-day conditions, disentangling literary studies from questions of nationality could also be a way of combatting the habitual neglect of authors whose works resist neat classification within a particular national literary context. According to Jay, it is precisely "[t]hese kinds of texts [that] are trans-

---

tional migrants has reached 258 million in 2017 (4). The same year has seen a record high of 68.5 million forcibly displaced individuals (UNHCR 2018, 2).

forming the scope of the national literatures to which they belong *and* pushing beyond national boundaries to imagine the global character of modern experience, contemporary culture, and the identities they produce" (9).

The transnational turn has also affected scholarship on the Beat Generation, a literary movement that has traditionally been studied within a predominantly American context (Grace and Skerl, 1). The past decade has seen a veritable boom in scholarship devoted to the Beat Generation's global impact and interconnections. Early publications concerned with the Beats' international artistic collaborations and their influence on artists across the globe include A. Robert Lee's *Modern American Counter Writing: Beats, Outriders, Ethnics* (2010), Nancy M. Grace and Jennie Skerl's *The Transnational Beat Generation* (2012), and a special issue of *Comparative American Studies* titled *The Beat Generation and Europe* (2013). In addition to highlighting the global connections of the most prominent Beat writers, these works also significantly extend the scope of Beat studies to include a larger number of artists. In *World Beats: Beat Generation Writing and the Worlding of U.S. Literature* (2016), Jimmy Fazzino considers the Beat Generation to be a "transnational literary movement par excellence" (4) and examines the significance of travel and international collaborative networks for the Beats, while a special issue of *CLCWeb: Comparative Literature and Culture* titled *Global Beat Studies* (2016) also features discussions on the various cultural influences manifested in their works. *The Routledge Handbook of International Beat Literature* (2018) and *Beat Literature in a Divided Europe* (2019) have looked beyond the US as a vantage point, with the latter in particular examining the reception and appropriation of Beat literature by European artists.

Movement, and particularly road travel, feature prominently in many Beat writers' works. As Deborah Paes de Barros (2009) states: "They drove their cars, not to impress or acquire or to arrive somewhere better, but to escape normative culture and to exist in the moment" (229). Rootlessness was championed as a necessary precondition in the quest for self-discovery (230). The journey became an end in itself, a way of arriving at a deeper understanding of oneself and the surrounding world. Even though weiss' work shares this exploratory, nonconformist spirit in texts such as "Compass" (published in *Can't Stop the Beat*) and "Big Sur" (1959), it often does so from the immigrant perspective of the refugee. In these works, the journey frequently becomes synonymous with a search for home. Her European roots and relations make her oeuvre stand out among those Beat writers who ventured beyond national borders in their lives and works. It is all the more surprising that weiss' name hardly ever receives more than passing reference in the above-mentioned publications, considering that much of her work can be seen as decidedly transnational, both with regard to its subject matter and the context of its creation.

Despite its centrality as a research focus, the meaning of *transnationalism* remains ambiguous and loosely defined (Pease, 4). In its most basic sense, the term refers to "multiple ties and interactions linking people or institutions across the borders of nation-states" (Vertovec, 447). Grace and Skerl offer several additional angles to the concept of transnationalism. It can be understood as "the loosening of cultural boundaries; [...] identities tied to multiple nation-states; [...] and an ongoing movement between two or more social spaces" (2). Movement, whether physical or metaphorical, is central to considerations of transnationalism. As Silvia Schultermandl and Şebnem Toplu (2010) have argued: "In this era of increasing global mobility, the nation-state can no longer serve as primary means of identification of selfhood" (11). Instead, "movement and motion become the primary metaphors of identity," meaning that "place ceases to be the sole representational parameter of identity and movement between places becomes the central space [for] a person's agency" (21). Experiences of movement as a means of self-definition can become a unifying link between people from different cultural backgrounds, a shared experience through which they are able to identify with one another (Vertovec, 450). Starting from this centrality of movement in transnational contexts, Stephen Greenblatt (2010) proposes looking at the literary representations of physical movements, the routes that enable them, as well as the places in which cultural exchange takes place in order to determine how these influence processes of identity formation in contexts of migration and cultural exchange (250 – 251).

weiss' life, shaped by displacement and both voluntary and involuntary movement, is reflected in an autobiographical poetic work that seeks to navigate the unstable ground between rootlessness and a search for home. The following analysis investigates how her work, as the result of such sometimes precarious circumstances, establishes a transnational perspective that increasingly eschews references to nationalities and, at times, conveys a postnational dismissal of nationality as a defining category of identification. In her poetry, she often portrays journeys of various kinds and the resulting encounters with people from different cultural backgrounds.

Peaceful Coexistence of Different Nationalities

weiss' hitchhiking trips across the United States are mentioned in the poem "One More Step West is the Sea," a reflection on five years of living in San Francisco (weiss 2018), which was written in 1957 and published in her first book, *Steps* (1958). In the poem, San Francisco is described as a multicultural transit point, the site of frequent departures and returns: "all the people are constantly crossing— / i've lived here five years / and only met two who were born here" (sec. 7).² The people's mobility is presented as the defining feature of the city which is portrayed as a space of wanderers in which neither origins nor destinations matter. At the same time, the shared experience of movement to the city functions as an implicit means of identification among its inhabitants. Here, having a migratory background is almost treated as the new norm while locals by birth are a rare occurrence. While building projects and the resulting destruction of historic architecture lead the speaker to muse on the ephemerality of the material world, she finds stability in the interpersonal encounters made on her journey: "but the people / oh the people / they can be so beautiful— / so ugly / so human-hungered / beautiful— / hitch-hiking across i found some / can't even begin to tell you how beautiful ..." (sec. 4). The speaker herself is identified as yet another wanderer who moved to the city from a different place. Her origins remain obscured when she states: "city, i have / come to you / from the farland... / my hands / are cupped" (sec. 5), suggesting that the name of her country of origin is insignificant. By thus eliminating all the connotations and preconceived notions that a nationality would inevitably conjure and that would separate her from others, the speaker refuses to let herself be defined by such attributes.

Encounters between people from different cultural backgrounds are further addressed in the long poem "Single Out," written one year later and published in 1978 in weiss' book of the same title. The poem consists of four sections, three of which are autobiographical. It constitutes weiss' first published engagement with her escape from Europe and the most explicit account of those experiences until the late 1970s. The first section titled "Little Girl" contains childhood memories of adapting to life in the United States and of the speaker's time in Vienna. A short exchange with another girl following the speaker's arrival in the US reveals ambiguous feelings about the new country: "how do you like america? / i laughed i like words / all the new words coming at me in the night / in bed awake listening to all the words / rushing at me all day." This playful and yet evasive response can be read as an indication that the speaker does not necessarily embrace the host country and its culture. *America* is just another word in the current of words which the newly arrived child picks up from around her. This emphasis on language foregrounds the act of communication and interpersonal connections. However, language also becomes associated with prejudice when her schoolmates call her "Nazi" because of her mother tongue: "the war was on / i spoke german / refugee a foreign word." This stereotyping of the

---

2 Unless indicated otherwise, none of the books by weiss discussed are paginated.

speakers of a language and, by extension, the citizens of a country, reveals a political mindset that does not distinguish between oppressors and victims. Such prejudices based on language and nationality are implicitly criticized through the Jewish speaker's subsequent bonding with a German girl and her family.

The speaker then begins to reminisce about Vienna where she had lived at her grandmother's boarding house. This boarding house, which frequently accommodated people from different countries, represents her first encounter with multiculturalism. Similar to the portrayal of San Francisco in "One More Step West is the Sea," it is depicted as a meeting point in which different cultural backgrounds do not play a role ("what a houseful changing & always the same"). The guests are united through their movement to the city from abroad. In this regard, the family's move from Berlin to Vienna in 1933 turns them into merely another group of travelers from a different place. Moreover, her grandmother's Hungarian roots are mentioned ("she was my hungary bridges of budapest"). Her grandmother embodies her connection to her ancestry from another country, rendering her a non-native to Vienna in more than one way. Crucially, her link with another place is not established through feelings of national affiliation but through interpersonal relations. Furthermore, the metaphor of the bridges connected to a country evokes notions of back-and-forth movement and exchange.

The atmosphere among the guests is described as harmonious. The Romanian students make music, while the Australian teachers and the Finnish couple bring along German copies of British and American books. The boarding house is thus presented as a counter-model to the previously mentioned xenophobia. In contrast to the speaker's American schoolmates, who let the political conflict dictate their opinion of her, "the chinese & the japanese medical students who were room-mates / wouldn't read the headlines that said war." Even though this section does not explicitly mention the political situation in the country that was then the so-called Austrofascist *Ständestaat*, its depiction of a multicultural microcosm shaped by peaceful coexistence and intercultural exchange also presents a sharp contrast to the rampant climate of racism and antisemitism that eventually forced the family to leave the country in 1938.

These political realities are the subject of the second section of "Single Out" titled "Incident." It details the family's failed attempt to escape to Switzerland, a few months after the country's integration into Nazi Germany: "october 1938 we had to flee vienna. / my grandmother hungarian boarding house / was wanted by a nazi official. / hungary was still out of nazi clutch & my grandmother hungarian. / we were austrian citizens – / my father, his mother's only son." This repeated juxtaposition of family relations with citizenship presents nationality as an unnatural and arbitrary construct that is used to decide who is persecuted and who is spared. In fact, at the time of their escape, weiss and her family were no longer citizens of Austria, as the country had been integrated into Nazi Germany under the name of *Ostmark* in March 1938. This mentioning of nationality masks yet another grim political reality: two and a half months after the "Anschluss," the so-called Nuremberg Race Laws were signed into law, stripping Jewish people of their citizenship (Burger, 146–147). The remark that they "had to flee vienna" also highlights an absence of identification with the country at large. Whereas other countries are mentioned by name (i.e. Hungary and Switzerland), this avoidance can be read as the speaker's (understandable) attempt to distance herself from said country.

Transnational Connections and Temporary Returns

Many of weiss' poems name cities rather than countries as people's places of origin and identification, with San Francisco and Vienna as the most frequently mentioned ones. Both cities feature prominently in weiss' poetic autobiography *Can't Stop the Beat*, especially in the poem "I Always Thought You Black." This poem is noteworthy as an autobiographical text in several ways. Rather than placing herself at the center of the narrative, weiss structured the work around a number of people who had played significant roles in her life. Furthermore, "I Always Thought You Black" takes a non-chronological approach to autobiography with interpersonal encounters causing the narrative to jump back and forth between various places in the United States and Europe that the speaker, at different times, has called home.

Following an account of the speaker's early days in the US, the cityscape of the new home in New York is contrasted with Vienna in a hint of nostalgia: "in the Bronx all the streets so even & clean & no green. / no one on the street. the houses straight up and down. / oh vienna my hand still on your grillwork. on your / balustrades" (12). References to the streets and to tactile memories of the city's architecture evoke associations with walking. Walking later connects the seemingly disparate places of San Francisco and Vienna: "san francisco. i'm home. a walking city. up & down / steps. vienna. not quite. a pang here & there" (25). Here, movement functions as a parallel connecting the old and the new home and thus creates a sense of belonging.[3] In 1952, San Francisco be-

---

[3] weiss has repeatedly stated that she felt particularly connected to San Francisco because it reminded her of Vienna. She sees both cities as walking cities and names this as one of the sim-

comes home when the speaker's hitchhiking trip leads her to the city and her driver, after taking her to the Bohemian quarter North Beach, "sez this is where you belong. right on" (24). In 1958, "home is a seven-room apartment. $60 a month. 1207 south van ness. heart of the mission. san francisco" (32). Then in 1967, the speaker makes her "home on vallejo street" (45). The way in which these specific locations are connected to the concept of home suggests that identification with a place and the sense of belonging associated with a home do not extend beyond the city to the state or country.

The text presents the concept of home as dynamic and subject to change. weiss has attributed her prolonged nomadic existence to a sense of restlessness and hesitation to "have roots again" which she connects to feelings of survivor's guilt ("Oral History Interview" part 2, 00:26:36). However, the casual way in which these changes of place are portrayed also allows for a more positive assessment of the idea of home as unstable. If home can be in many different places, it does not depend on identification with a specific country or culture. It also suggests that home can be in multiple places at once and, as the references to Vienna and other European cities show, is not restricted to a single nation.

Movement between places is a constant reality and thematically connects the poems in *Can't Stop the Beat*. The book's dedication to "all the travelers who trust the bend in the road" (n. pag.) further emphasizes this theme. Mary Carden (2018) reads this preoccupation with movement as representing a metaphorical journey of self-discovery (85). She points out:

> Roads and paths are common symbols for the trajectories of human lives, and readers are accustomed to finding processes of self-definition represented as quests or journeys [...]. Selves-in-progress have much in common with travelers seeking their way, encountering the unexpected, accommodating changing circumstances, and amending what they think they know about themselves and about their social and cultural contexts. (188)

Carden here harks back to the aforementioned Beat theme of the journey as self-discovery, and even though she acknowledges the ways in which journeys can change a person's perspective on their own culture, she stops short of pointing out the substantial shaping influence of weiss' autobiographical journeys on the worldview reflected in her work. Instead, she goes on to state that "weiss writes autobiographical prose poetry infused with motion and elasticity [...] Describing a self in continuous process [...], weiss blurs boundaries of gender, race, and sexuality to offer [...] multiple and open possibilities for identity" (190). Her analysis

---

ilarities between them which made her choose San Francisco as her adopted home of twenty years (weiss 1993, "Oral History Interview" part 2, 00:29:35; weiss 2018).

identifies in weiss' work a subversion of gender stereotypes that have proven particularly pervasive within the context of largely male-centered Beat writing. From a gender perspective, her independent movements (hitchhiking by herself in the 1950s, walking alone at night, going on road trips as a woman, etc.) can indeed be read as a challenge to the Beat trope of the lone male traveler. Her work shares this quality with the writings of other female Beat artists, e.g. Diane di Prima's *Memoirs of a Beatnik* (1969), Janine Pommy Vega's *Tracking the Serpent: Journeys into Four Continents* (1997), and Hettie Jones's *Drive* (1998), which similarly portray female travelers.

However, weiss' autobiographical work, referencing her refugee past and life in Europe, also offers an alternative perspective on the Beat trope of the journey as exploratory and freeing by evoking associations with uprootedness and a search for home. In her work, movement is a constant reality, yet it is not necessarily always presented as desirable. Her poetic representations of movement extend beyond well-worn autobiographical metaphors. *Can't Stop the Beat*, a book that, not least through its title, stresses weiss' role in the Beat movement, at the same time introduces her European background and thereby calls into question a (by now outdated) idea of the Beat movement as "quintessentially 'American'" (Fazzino, 4). Through her self-representation as a "Beat poet," she not only highlights the presence of a female artist but also of an immigrant. By foregrounding autobiographical movements, she acknowledges their influence on her worldview and her mobile conceptualization of home that transcends national borders. The text does not contain any extensive passages about the time she spent in Europe. Nevertheless, the presence of vague allusions to a non-American past is crucial, as they raise questions regarding the speaker's origins but leave them unanswered, thereby challenging their relevance. While undoubtedly a shaping influence for it to merit repeated mentioning, weiss' time in Vienna is not presented in a way that paints her as a European in exile.

Grace and Skerl's observation that weiss, through her non-US background, carried with her "an identity that bridged both nation-of-birth and adopted country" (3) acknowledges the transnational perspective that makes her work unique within Beat writing. However, rather than her birthplace Berlin, which only receives passing reference in her poetry, it is Vienna, where weiss lived from the age of five to ten, which has left a lasting impression on her. In fact, the connection runs so deep that, to this day, she calls it her hometown (weiss 2018). In 1998, five years after writing "I Always Thought You Black," weiss would return to Vienna for the first time in fifty years (Neundlinger, 67). What was only her second visit to the city since her escape proved consequential for her subsequent career.[4] A performance at the "Beat Generation Fest" in Prague in April 1998 marked the beginning of a productive creative period that led to a number of artistic projects and collaborations across Europe (Zobel, 31). Her activities during her frequent visits to Vienna over the subsequent fifteen years included teaching a poetry workshop at the Schule für Dichtung (Vienna Poetry School) in September 1998 (Neundlinger, 67), numerous jazz & poetry performances, the publication of a total of six bilingual books with Austrian publishers, as well as a theatre performance of three of her plays in October 2006 ("ruth weiss-Festival in Wien"). Evidently, all these have played a decisive role in weiss' career as, for the first time, her work became known to an Austrian audience.

In the poem "A Promise Kept" (2013), weiss pays homage to this creative period, as well as to her ongoing connection to Vienna. Despite this fondness for the city, weiss never intended to return to Vienna permanently (weiss 2018). Instead, "A Promise Kept" posits the return as a repeated action that fosters exchange. The poem is written as an apostrophe to the mythological figure of Prometheus: "oh PROMETHEUS / help me to return to wien / again + again / to walk vienna woods / to walk vienna boulevards / gather chestnuts in the parks / balustrades + balconies / poetry + jazz / free from limitations // oh vienna / of water pure + air to breathe / let me return / again + again" (1–13). As in "I Always Thought You Black," the city is explored on foot. The act of walking suggests the fleetingness of her stays in Vienna which are presented as mere stopovers. Rather than settling down, the speaker remains in constant motion. Walking also presents a dynamic counter image to a life fixed in one place and conveys a sense of mobility as the prerequisite for a transnational existence. Instead of reinforcing any national affiliations, the repeated return enables her to maintain a connection to the city "free from limitations." The return as self-chosen movement allows her to actively position herself between two places and to re-establish and maintain a relationship with the former home. Like "Single Out," this poem, as well as most of weiss' other "return poems,"[5] speaks of a return to *Vienna* as opposed to *Austria*. As with the majority of her poems referencing the capital city of Austria, the country itself remains unnamed, expressing an emotional connection to the city as a stable reference point while suggesting a lack of identification with the country as a whole.

---

4 She briefly visited Vienna in 1947 while attending college in Switzerland ("Oral History Interview" part 1, 01:59:00).
5 See also "As the Wheel Turns" (1998) and "Torch-Song for Prometheus" (1999, publ. in *Full Circle/Ein Kreis vollendet sich* [2002]).

Absence of Nationality

While nationalities and specific places were still present in weiss' earlier poems, over time, her work presents localities in increasingly abstract ways to the point where fixed location markers are dismissed altogether. This tendency is most drastically realized in the book *Desert Journal* (1977). The eponymous desert is an imagined mental landscape ("one invents a place to be— / a place that is / precisely what one is" ["Tenth Day"]) and the forty poems forming the collection have been read by Nancy M. Grace (2004) as representing "an internal journey towards self-discovery" (60). This "internal desert," as weiss also describes it (*Desert Journal* 2012, 207), is at times presented as disorienting, lacking clear points of reference: "so this is the desert [...] there are no stops / and no horizon / there is no path / [...] there is no up / there is no on / there is not even down [...] there is no old / no feet to walk / no hands to hold / and nothing in-between" ("Twenty-Eighth Day"). As Grace sees it: "[H]er desert is positioned as a landscape only vaguely recognizable on a human scale [...]. Immersed in this imaginative space, weiss succeeded in conveying elements of personal and cultural history in a poem cycle that rejects tribalism, confession, and identity politics" (60). The speaker's sense of identity can thus not be tied to any particular place: "one changes rooms / one changes buildings / one changes cities / one changes countries / one changes continents / planets // it is still the same universe / one's own" ("Fifth Day"). This statement reiterates the aforementioned irrelevance of place as a source of identification and presents the individual's sense of self as continuous in the face of movement between places. Moreover, the absence of a recognizable location makes identification with a place impossible.

Carden reads the work's emphasis on the journey as opposed to a recognizable space as an attempt at "constitut[ing] identity in/as movement" (85). This paradoxical grounding in motion is most pronounced in the following lines: "if even one petal moves / it will CATAPULT / me out of the desert / OUT OF THE DESERT! // but this is my home— / this wandering" ("Twenty-First Day"). Within the imaginary space of the desert, the speaker thus presents wandering as a source of identification. Carden interprets this decontextualization as a critique of "models of identity [...] that demand and perpetuate separation and exclusion as ordering principles" (88). In this regard, the absence of concrete location markers and a recognizable space can be read as calling into question the relevance of nationalities for human coexistence. Or, as Carden further states: "Setting aside all the markers cultures devise to make distinctions between selves, weiss' poems represent self-discovery as mutable, multiple, and ongoing, a journey that intersects with the journeys of other selves. [...] weiss' poetry consistently breaks down borders that restrict free exchanges of 'you' and 'all'" (89). Accordingly, on the "Tenth Day," the speaker blurs distinctions between the self and humanity at large: "what is that face / of a whole human race / that gazes through one's eyes?"

This statement presents an egalitarian image of humanity that can be connected to weiss' evolving view of the concept of nationality as essentially meaningless (weiss 2018). The issue is also featured in her interview for the San Francisco *Holocaust Oral History Project* which she gave in two sessions in May and July 1993. In the five-hour interview, she elaborates on her escape from Vienna, her process of adapting to life in the United States, and the effect of these experiences on her worldview. In a personal interview, weiss explained that she participated in the *Oral History Project* on the condition that she would talk about people from Germany and Austria who rejected National Socialist ideology and had helped her and her parents in different ways, to counter sweeping condemnations of citizens of those countries as perpetrators. She also acknowledged the strong impact that this interview had had on her, as it made her realize how profoundly her escape and the resulting feelings of survivor's guilt had affected her (weiss 2018). This change in perspective also influenced her later work by triggering a new engagement with her past and the importance of active remembering.

The interview is the subject of the poem "Her Number Was Not Called" (published in the bilingual book *No Dancing Aloud/Lautes Tanzen nicht erlaubt* [2006]), perhaps weiss' most explicit grappling with the Holocaust and the guilt of the survivor. The first part of the poem refers to the speaker's traumatic flight from Vienna to evade Nazi persecution, foregrounding the movement of escape: "she sails away [...] salt of the sea / mixes with tears / that do not return" (6). The interview situation offers catharsis in the act of remembering: "we are here to gather the stories from the HOLOCAUST / my story? / my number was not called // i was not there to be given a number [...] but you are here / yes i am here / i left with mother & father— / [...] all the others— / into gas & smoke" (6 – 8). Locations remain unnamed in the poem and are only referenced through the adverbs *here* and *there*. While the subject matter unambiguously identifies *there* as Nazi Germany and would seem to render an explicit mentioning of the place redundant, the complete absence of geographical location markers points towards the universal validity of the poem's message. Furthermore, the speaker's parents are called "mother & father" instead of, as is usually the case in weiss' poems, "mutti" and "papa". The use of the English terms conceals the speaker's origins which otherwise would have become explicit. In this manner, the speaker highlights the irrelevance of precisely those national categories and their fanatic glorification that are the source of the suffering portrayed in the poem.

Conclusion

weiss' work, due to its thematic engagement with American and European settings, multilingualism, and publication in different countries, evades classification within a single national literary context. A consideration of a body of work such as hers calls into question traditional frameworks of literary analysis within national boundaries. Furthermore, it can show how trans- and postnational perspectives emerge as attempts at coping with the potentially disruptive effects of the experience of displacement on an individual's identity.

To conclude this chapter, it can be said that weiss' poetry rejects the notion of nationality as a defining category in favor of a worldview in which human coexistence is grounded in experiences and interpersonal relationships rather than questions of origin. Her early encounter with the devastating potential of nationalism has instilled in her a deep skepticism of all forms of patriotism that is also reflected in her work (weiss 2018). Over time, weiss' preoccupation with what unites and divides people led to her engagement with locality and, by extension, nationality, in varying degrees of abstraction. In her poetry, specific places such as cities are frequently used as the only geographical points of orientation. Furthermore, these are presented as spaces in which people from different countries come into contact with each other and often form transcultural connections. Ultimately, it does not matter whether these conversations take place in Berlin, San Francisco, or Vienna. In "One More Step West is the Sea" and "Single Out," encounters between people take place on the move. Movement is presented as a unifying link between people, calling into question the concept of *home* as fixed in a particular place or linked to a nationality.

Not only do people transgress national borders through their physical movements between cities, but the cities themselves also become linked through their potential for movement (both Vienna and San Francisco are described as "walking cities"). In "I Always Thought You Black" and "A Promise Kept," autobiographical poems that highlight weiss' lasting connection to the city of Vienna, a transnational perspective is established that combines multiple cultural influences without privileging one over the other. By reimagining the concept of home as dynamic and shaped by movement, the poems' speakers maintain ties to multiple places while resisting confinement to one specific location.

References to nationalities are conspicuously absent from many of weiss' later poems. In connection with weiss' own statements, these could be read as an attempt at postnational identity formation, showing national categories as, at best, irrelevant for and, at worst, detrimental to peaceful human coexistence. *Desert Journal* presents a thought experiment that radically questions the need for national categories as a means of identification by eliminating recognizable locations altogether. In the poem "Her Number Was Not Called," the complete absence of the names of countries and even cities constitutes a resolute rejection of nationalist ideologies and presents the construct of nationality as insignificant. As she sums it up on the "Tenth Day" of *Desert Journal:* "the place is unimportant / until called / given a name / put into a setting / become responsibility."

Works Cited

Burger, Hannelore. *Heimatrecht und Staatsbürgerschaft österreichischer Juden: Vom Ende des 18. Jahrhunderts bis in die Gegenwart*. Wien: Böhlau Verlag, 2014.

Carden, Mary Paniccia. *Women Writers of the Beat Era: Autobiography and Intertextuality*. Charlottesville: U of Virginia P, 2018.

di Prima, Diane. *Memoirs of a Beatnik*. Paris: Olympia Press, 1969.

Edwards, Brian T., and Dilip Parameshwar Gaonkar, eds. *Globalizing American Studies*. Chicago: U of Chicago P, 2010.

Fazzino, Jimmy. *World Beats: Beat Generation Writing and the Worlding of U.S. Literature*. Hanover: Dartmouth College Press, 2016.

Fisher Fishkin, Shelley. "Crossroads of Cultures: The Transnational Turn in American Studies: Presidential Address to the American Studies Association, November 12, 2004." *American Quarterly*, vol. 57, no. 1, 2005, pp. 17–57. Web: doi:10.1353/aq.2005.0004. Accessed 20 Feb. 2019.

Giles, Paul. *Virtual Americas: Transnational Fictions and the Transatlantic Imaginary*. Durham: Duke UP, 2002.

Goyal, Yogita, ed. *The Cambridge Companion to Transnational American Literature*. Cambridge: Cambridge UP, 2017.

Grace, Nancy M. "ruth weiss's *Desert Journal:* A Modern-Beat-Pomo Performance." *Reconstructing the Beats*, edited by Jennie Skerl, New York: Palgrave Macmillan, 2004, pp. 57–71.

Grace, Nancy M., and Jennie Skerl, eds. "Introduction to Transnational Beat: Global Poetics in a Postmodern World." *The Transnational Beat Generation*, New York: Palgrave Macmillan, 2012, pp. 1–11.

Greenblatt, Stephen. "A Mobility Studies Manifesto." *Cultural Mobility, A Manifesto*, edited by Greenblatt et al., Cambridge: Cambridge UP, 2010, pp. 250–253.

Harris, Oliver, and Polina Mackay, eds. *Global Beat Studies*. Spec. issue of *CLCWeb: Comparative Literature and Culture*, vol. 18, no. 5, 2016. Web: doi:10.7771/1481–4374.2980. Accessed 20 Feb. 2019.

Jay, Paul. "Beyond Discipline? Globalization and the Future of English." *PMLA*, vol. 116, no. 1, 2001, pp. 32–47. Web: www.jstor.org/stable/463639. Accessed 20 Feb. 2019.

Jay, Paul. *Global Matters: The Transnational Turn in Literary Studies*. Ithaca: Cornell UP, 2010.
Jones, Hettie. *Drive*. New York: Hanging Loose Press, 1998.
Lee, A. Robert. *Modern American Counter Writing: Beats, Outriders, Ethnics*. New York: Routledge, 2010.
Lee, A. Robert ed. *The Routledge Handbook of International Beat Literature*. New York: Routledge, 2018.
Mackay, Polina, and Chad Weidner, eds. *The Beat Generation and Europe*. Spec. issue of *Comparative American Studies*, vol. 11, no. 3, 2013. Web: doi:10.1179/1477570013Z.00000000042. Accessed 20 Feb. 2019.
Neundlinger, Helmut. "Nachtvogel singt." *Datum*, vol. 1, 2007, pp. 64–67.
Paes de Barros, Deborah. "Driving That Highway to Consciousness: Late Twentieth-Century American Travel Literature." *The Cambridge Companion to American Travel Writing*, edited by Alfred Bendixen, Cambridge: Cambridge UP, 2009, pp. 228–243.
Pease, Donald E. "Introduction: Re-Mapping the Transnational Turn." *Re-Framing the Transnational Turn in American Studies*, edited by Winfried Fluck, Donald E. Pease, and John Carlos Rowe, Hanover: Dartmouth College Press, 2011, pp. 1–46.
Pommy Vega, Janine. *Tracking the Serpent: Journeys into Four Continents*. San Francisco: City Lights, 1997.
"Ruth Weiss-Festival in Wien." *Der Standard* 10 Oct. 2006. Web. https://derstandard.at/2605839/Ruth-Weiss-Festival-in-Wien. Accessed 20 Feb. 2019.
Schultermandl, Silvia, and Şebnem Toplu, eds. Introduction. *A Fluid Sense of Self: The Politics of Transnational Identity*, Vienna: LIT Verlag, 2010, pp. 11–24.
United Nations. "Department of Economic and Social Affairs, Population Division." *International Migration Report 2017: Highlights*. New York: United Nations, 2017.
United Nations. "High Commissioner for Refugees." *Global Trends: Forced Displacement in 2017*. Geneva: UNHCR, 2018.
Veivo, Harri, Petra James, and Dorota Walczak-Delanois, eds. *Beat Literature in a Divided Europe*. Leiden: Brill Rodopi, 2019.
Vertovec, Steven. "Conceiving and Researching Transnationalism." *Ethnic and Racial Studies*, vol. 22, no. 2, 1999, pp. 447–462. Web. doi:10.1080/014198799329558. Accessed 20 Feb. 2019.
weiss, ruth. *Steps*. San Francisco: Ellis Press, 1958..
weiss, ruth. "Big Sur." *Beatitude* 3 (23 May 1959): n. pag., 1959.
weiss, ruth. *Desert Journal*. Boston: Good Gay Poets, 1977.
weiss, ruth. *Single Out*. Mill Valley: D'Aurora Press, 1978.
weiss, ruth. *Oral History Interview with ruth weiss*. 14 May and 22 July 1993. The Bay Area Holocaust Oral History Project. *United States Holocaust Memorial Museum*, 1993. Web: https://collections.ushmm.org/search/catalog/irn509624. Accessed 20 Feb. 2019.
weiss, ruth. "As the Wheel Turns." MS. Bancroft Library, University of California, Berkeley. BANC MSS 2010/153. Ruth Weiss collection of poetry broadsides and cards, 1975–2016. Oversized Box 1. Folder 3, 1998.
weiss, ruth. *No Dancing Aloud/Lautes Tanzen nicht erlaubt*. Bilingual English/German. Trans. Horst Spandler. Vienna: Edition Exil, 2006.
weiss, ruth. *Can't Stop the Beat: The Life and Words of a Beat Poet*. Studio City [Los Angeles]: Divine Arts, 2011.
weiss, ruth. *Desert Journal*. 1977. New Orleans: Trembling Pillow Press, 2012.

weiss, ruth. *Full Circle/Ein Kreis vollendet sich*. 2002. 2nd rev. ed. Bilingual English/German. Trans. Christian Loidl. Vienna: Edition Exil, 2012.

weiss, ruth. "A Promise Kept." October 2013. MS. Bancroft Library, University of California, Berkeley. BANC MSS 2010/153. Ruth Weiss collection of poetry broadsides and cards, 1975–2016. Oversized Box 1. Folder 2, 2013.

weiss, ruth. Personal interview by Stefanie Pointl. 27 July 2018. San Francisco, California, 2018.

Welsch, Wolfgang. "Transculturality – The Puzzling Form of Cultures Today." *Spaces of Culture: City, Nation, World*, edited by Mike Featherstone and Scott Lash, London: Sage, 1999, pp. 194–213.

Zobel, Greg. "Ruth Weiss: Making A Deeper Groove in Europe." *Kerouac Connection*, vol. 29, 1998, pp. 31–33.

Estíbaliz Encarnación-Pinedo
"How real is i?": Gender and Poetics in ruth weiss

Fig. 1: ruth weiss in 1973 (photograph by Scott Runyon)

Introduction

> "I'd like here to declare an enlightened poetics, an androgynous poetics, a poetics defined by your primal energy not by a heterosexist world that must measure every word, every act against itself." (Anne Waldman, "Feminafesto", 1994, 145)

> "we are of man and woman / and having named us / both will still / use / us as battlefield // we scream / to find our balance" (ruth weiss, *Blue in Green*, 10)

Dressed in black and framed under a huge, broad-brimmed hat, ruth weiss gives the cool beatnik chick a distinctive twist. The thick black eye shadow and dark eyeliner grant weiss an out-of-this-world, sorcerer-like look as she poses for Scott Runyon – the photographer who often documented the Haight-Ashbury psychedelic drag theater group The Cockettes in Steven Arnold's films. Much too intense to be an embodiment of the "[m]ulberry-eyed girls in black stockings" recorded in Bob Kaufman's sharp "Bagel Shop Jazz" (2019, 11), weiss sits more comfortably with the image of the hag. Now a well-established symbol of female transgression (Firestone 1970; Daly 1978), the hag has been celebrated by other poets associated with the Beat Generation such as Anne Waldman or Diane di Prima[1]. As someone who has "thrown off shackles of mean expectation [and] could finally manifest beyond 'girl', 'wife,' 'mother'" (1994, 142), Waldman restores in her essay "Feminafesto" the disruptive ambivalence of the hag to introduce a "poetics of transformation beyond gender" (145), a freer creative space where the poem is defined by an energy beyond the social norms that regulate both women and men.

weiss, who has only truly felt comfortable with the label of poet – or better yet "Jazz Poet" (2002, 166) – and has herself avoided the constrains attached to the female sex in the Cold War period, constructs throughout her body of work a poetic stance which systematically disrupts and dissolves gender constructs and which, as Mary Paniccia Carden notes, "critiques and reformulates models of identity based in binary opposition, in either/or organizations of being that demand and perpetuate separation and exclusion as ordering principles" (2018, 88). In poetry collections such as *Blue in Green* (1960), *Light and Other Poems* (1976), *Desert Journal* (1977), in plays such as *Figs* (1960), as well as in her underground film *The Brink* (1961), weiss shows a preoccupation with themes of identity construction and expression which often stem from the opposition of normative gender dichotomies that situate women and men in battle. Much like the archetypal He and She of *The Brink*, "man" and "woman" in "Blue in Green" – the poem included in the epigraph of this chapter – find themselves struggling to find a balance between their own identity and self-expression and the categories they are placed into.

In the above-mentioned works, as well as in her overall oeuvre, weiss complicates and blurs established categorizations through which she documents both the struggle and the balance, the exclusion and the dissolution of the (de)gendered selves that inhabit her work. To study the ways in which weiss resolves these tensions, this chapter analyzes the thematic traits as well as the stylistic choices that allow her to write *beyond gender* and avoid the "norm that assumes a dominant note subordinating, mistreating, excluding any possibility" (Waldman 1994, 145). First, it situates weiss and her work in the heteronormative,

---

[1] See Anne Waldman's "Hag of Beare (Caillech Berri)" in *The Iovis Trilogy* (2011) or Diane di Prima's "Vision of the Hag, devoured" in *Loba* (1998).

highly gendered, era of the American post war years. Secondly, it delves into weiss' poetics and aesthetics in relation to the dissolution, or at least the contestation, of the gendered poetic "I".

Whose normativity? weiss and the liberation of gender

The American post-war period, with its aura of paranoia and societal pressure to conform, has been widely documented. While it is true that Beats emerged in a period where "gender norms were highly rigidified" (Davidson 2004, 161) and where both sexes "were under intense social pressure to enter the fray and resolve it by 'settling down'" (Ehrenreich 1983, 2), women often found bigger and more threatening obstacles if they deviated from the norm. Nevertheless, while Goodman (in)famously neglected any form of female rebellion in his 1960 sociological study *Growing Up Absurd*, others have drawn attention to relaxation and availability of alternative gender roles for women in bohemian circles, even if they were mostly suppressed or ignored (see Breines 1994). Indeed, many of the so-called women of the Beat Generation, as Maria Damon observes, "occupied positions beyond those of, on the one hand, sex objects and, on the other, Momist matriarchal tyrants hellbent on turning apron strings into straightjackets for their male partners and progeny" (1995, 146). After reading *On the Road*, author Joyce Johnson saw herself represented not in the women in the novel – who appeared to her boring and conformist – but in Sal Paradise, who was "passionately impatient with the status quo" (1999, 46). Acting out on that passion, so manifestly assumed for the masculine sex, placed many of these women as kinds of gender transgressors or agitators. Reflecting back on the unconventional lives they had chosen, Johnson quotes from poet and friend Hettie Jones to describe the unorthodox position they occupied in society: "We shared what was most important to us [...] common assumptions about our uncommon lives. We lived outside, as if. As if we were men? As If we were newer, freer versions of ourselves?" (47).

When the twenty-something, short-green-hair ruth weiss hitchhiked her way into San Francisco in 1952, she seems to have been instantly and effortlessly assimilated by the countercultural sphere. As if organically assumed by the bohemian fabric of the city, there was little doubt that weiss would live outside conventions of gender normativity – "my last ride drops me on broadway & columbus. north / beach. sez this is where you belong. right on." (2011, 24) In keeping with the ease of these lines, weiss rarely uses her work to denounce or expose gender disparities or an unbalanced access to art and poetry; unlike poets like Anne Waldman, Diane di Prima or Joanne Kyger – who have more overtly dealt with gender issues – weiss has favored a poetic stance in which gender, of the speaker or the subjects in her poems, either assumes a secondary position, or is increasingly diluted until being barely represented. Yet, this image of natural and painless belonging was not ignorant of the gender dichotomies that prevailed even within bohemian and artistic circles. As Zida Borcich writes:

> Out there, where [weiss] had been hitching rides and meeting the median population for a year or so, a Father-Knows-Best ethos prevailed: work your way up the corporate ladder, climb the social staircase, have a couple of kids, buy a house, save for retirement, work at a job with benefits and pension (whether it interested you or not), dress suitably, speak properly, believe appropriately, act in a manner prescribed by prevailing mores and entertainment industry's ideas of correctness, and, above all, consume. (2013, 4)

Not oblivious to the culture that she participated in, weiss fought many of these dictates in her own life, many of which, even if not directly tackled in her work, were clearly based on her gender. For instance, while poets like Diane di Prima or Hettie Jones have addressed issues of motherhood and poetry in their work – the former even setting motherhood as essential for her poetic vision[2] – weiss early on rejected motherhood as an appropriate role for her as it felt antithetical to being a poet. In her own words: "I always wrote. Already knew as a child I would be a poet. That's why I didn't want to have children. I wanted to be free. Wrote my first poem when I was five years old." (2002, 166) As this statement suggests, the pressure of the previously-mentioned social mandates, as well as the implications they entailed for both sexes, do have an impact in weiss' life and poetics even if, in light of the above, her poetics favors a unifying, collective, vision of humankind. This is the case, for instance, in "One More Step West Is the Sea", the long narrative poem that opens her first collection, *Steps* (1958). In this poem weiss describes a San Francisco "of the spread-around hills where the action shifts while its geometrical center is still an outpost" (n.p.), establishing a battlefield connection which anticipates a threat:

> silly city! The webbed monster is clawing you
> he will devour you when he has ceased playing.

---

[2] See *Recollections of my Life as a Woman* (di Prima 2001, 161–162). Waldman similarly links her own experience of motherhood with the development of her poetic career: "When I came into power as a writer, and I think this had to do with becoming a mother as well, I could say outrageous things, could proclaim my 'endometrium shedding.' Could manifest the 'crack in the world.' I shouted, 'You men who came out of my belly, out of my world, BACK OFF!'" (1994, 144).

> the tower of babel will not be built again.
> but there are other symbols that must first
> come concrete. (n.p.)

Very reminiscent of Allen Ginsberg's Moloch in "Howl" (1955), who was "sphinx of cement and aluminum" (54) and "whose blood is running money!" (54), weiss' "webbed monster" is given the intangible and multi-layered form of modern civilization, a threat almost accepted by an oblivious city that is progressively transforming into an alienated space – "you have invited the webbed monster to connect you he has a cement and steel-greedy eye" (1958, n.p.). Like Ginsberg, weiss links her monster to recurrent threats of the era such as the rat-race, sexual inhibition, or the fear of nuclear annihilation – "only paper is king / kong gone / down the river / of again / because in sex-fear / orgasm-fear / we must build better and bigger bombs" (n.p.). While some of these issues had different implications for women and men, in this poem weiss uses a personal and yet un-gendered speaker through which she calls for the unifying voice of humankind that emerges "so ugly / so human-hungered" (n.p.) and yet "beautiful—" (n.p.). At the end of the poem, an ethereal, non-body-specific love and the speaker's cupped hands to symbolize acceptance become the city – and therefore its inhabitants' – salvation. This peaceful and almost celestial love contrasts with other works in which the female and male gender are placed against each other, often coming into a conflict that seems to be at the core of their being. Part of this conflict, as we can see in *Figs* (1960) – weiss' first play – stems from the different labels attached to them, either externally, mutually or individually. In this play, weiss refashions Adam and Eve into an even more archetypical embodiment of the gender dichotomy: "the first he and she" (2006, 106). In a highly metanarrative dialogue, HE and SHE are well aware of their own power to name, and therefore label, the things they see around them:

> HE   would you care for a fig
>
> SHE   it's a fig to me
>   besides, what makes figs, figs
>
> HE   *thoughtfully* what would a fig be if it wasn't a fig (110)

The answer to this non-question, and the liberating power of its inconclusiveness, lies at the heart of the couple's politics of identity as they come to realize that their power to label things includes themselves too. Perhaps wanting to be that fig which is not a fig, SHE fights the role she has been assigned by her partner and uses language to anticipate and debunk with irony any sort of scripted or normative relationship between the two:

HE	but you're a thousand women
SHE	1001
HE	all right, 1001, you are—
SHE	your sunshine
HE	yes my sun my moon my star
SHE	*peevishly* always yours
HE	but you want it that way
SHE	don't you ever think of others *dreamily* ME ME ME (114)

This tension is mirrored and extended in another set of "he" and "she"; namely, in weiss' *The Brink* (1961), the film she directed and the script of which is a collage of parts of the narrative poem "The Brink", published in the 1978 collection *Single Out*, and excerpts of different poems written in the late 1950s and early 1960s. As if directly picking up the thread from *Figs* – lines of which are actually used in the film – *The Brink* continues the investigation of identity construction in the context of gender opposition that arises from the couple's love story. In one of the first scenes of the film, He and She sit across each other in a cafeteria while a mirror doubles up their image as if anticipating the film's preoccupation with issues of identity, individuality, and subordination between the sexes. It is in this context that the narrator directly questions the roles He and She are to fulfill: "are you my day, my dawn, my sun? am I to walk with you? Am I to talk to you?". These questions, as the film progresses, become more and more relevant for both He and She, as the couple's cyclical refusal of, as well as longing for, connection leads them to question their individuality and their self-expression in association with the other[3]. Hence, the mirroring of the line "go to the round house he can't corner you there" / "go to sea she can't corner you

---

[3] This theme is also found in weiss' play *The Thirteenth Witch* (1981), a play that shares with *Figs* a preoccupation with its own form and with the source of creativity. To tackle this subject, weiss rewrites the Grimm Brother's sleeping beauty fairytale, setting it, as weiss puts it, "into its opposite context." (2006, 122) Macumbre, the thirteenth witch who was not invited to the princess's feast in the original tale, does not seek revenge in weiss' play – by cursing the princess with a hundred-year sleep – but rather wishes to give her the gift of love and creativity. Unlike the other witches' "temporal" (122) gifts – such as beauty or riches – Macumbre's gift represents "transformation in a rigid-rule world." (122) As part of this rigid world, there are "all the false / claiming love as their goal / but only really scrambling for possession" (184).

there" foregrounds the poet's insistence of gender balance or equality. In this sense, She's no-corners, round, house mirrors the rejection of normative domesticity also shared by He's boundless and borderless sea. That is to say, both genders, though maybe facing different obstacles, are subject to the same oppression.

As such, weiss uses the archetypical "he" and "she" – embodiments of the gender divide the poet often contests in her work – not to denounce women's subordination to a patriarchal, all-governing, power, but to stress the dangers embedded in the categories, labels, and fixed molds which both men and women are put into. In the film, it is not until some of these threats are resolved – for instance the baby carriage that blocks their way or the suburban houses that echo Norman Mailer's slow death by conformity (1957) – that He and She are finally free to create a space in which they can both be themselves individually and with one another. This peaceful reconciliation of the sexes happens, not coincidentally, at the margins of society, in the debris and ruins where the couple can begin to build up a new relationship, a new way of interacting with each other.

As these examples attest to, weiss' poetry shows an understanding of the complex and often oppressive gendered discourses that affect both women and men and which – more interestingly for the poet – negatively affect their creativity. Perhaps to free language from the body it channels through, this thematic interest often turns in weiss' work into a de-genderized poetics in which, as in Waldman's creative "Feminafesto," the poet's "body is an extension of energy" (1994, 145); an extension of the sex chromosomes weiss purposefully mixes up in the poem "Something Current"[4] (1978) and which can be realized through the creative power of art. Using language as the debris through which He and She rebuild their connection, weiss often uses word play, syntax and the juxtaposition of images to symbolically break down the barriers and the categories that keep women and men, as well as poetry itself, contained. The next section delves into the dissolution of the poetic "I" as one of the ways in which weiss' poetry uses language to break down these categories.

---

[4] "this is an i poem / this is a look straight into the eye poem / this is a you poem / you / we / W X Y Z" (1978, n.p.)

Who's/whose writing? weiss and the liberation of the poetic voice

In "2009", a poem included in *A Parallel Planet of People and Places* (2012a), weiss urges the reader to "set sail from the I-LAND" (61) and "be a live girl & boy / be a live boy & girl" (61). This move, closely linked to her poetic stance, is in the poem a call for the dissolution of the strings attached to the gender-specific "I", a strategy designed to remove barriers – "cast out negation" (61) – and expand the creative energy – "make space / give air for the fire from the heart" (61). The resistance to gendering labels analyzed in the previous section finds its ultimate transformation in weiss' dismissal of an over-ruling and domineering poetic "I". In this light, the poet's own resolution to spell her name in lower case – a "rebellion against law and order" (qtd. in Grace and Johnson, 2004, 69) – becomes an aesthetic stance in which the lower "i" becomes "a contingent entity and not a controlling one – a sign among signs" (Whaley 2004, 51). Well aware of the danger of letting the ego control the poem, weiss has favored a vision of poetry by which the poet becomes the instrument[5], the medium through which poetry happens. For example, in collections such as *South Pacific* (1959) or in the poem "Africa" (*Can't Stop the Beat*) the idea of a received poem, in which the poet acts as intermediary, is explicitly addressed. This highlights the notion that, as Nancy Grace has noted, weiss "defines language as a free-flowing force moving outward from the unconscious toward self and other, a phenomenon grounded in her belief that 'language is sacred, therefore dangerous. To be used with care'" (2004, 61). This idea of a language freed from constrains imposed by various domineering discourses – from the self-imposing "I" to a strict understanding of art and poetry – is present in the two collections analyzed in this section: *Desert Journal* (1977) and *Light and Other Poems* (1976).

Perhaps weiss' most ambitious examination of the poetic "I" is her long poem cycle, and to many masterpiece, *Desert Journal*. Nancy Grace has described the speaker's voice in these poems as "a de-personalized, de-sexualized, de-gendered, and de-historicized voice operating much like a mechanical recording device." (2004, 62) Much of this process of *de-ing* the poem has to do with freeing language by eliminating the external elements of control, being the personal, sexual and gendered poetic voice the first one to go. Transformed into the "kan-

---

5 weiss' own portrait-poem in "For these Women of the Beat" reads "the beat the beat the beat / the poet to be instrument / to keep the instrument clear / sound a new view of matter / or an ancient one regained" (2012a, 29). These verses also bring to mind weiss' use of her own voice as instrument when performing with live jazz musicians.

garoo-bird" that is "either sex / or both" (weiss 1977, n.p. ["First Day"]) in the first poem, the speaker loses the authority, and therefore gains the creative freedom, to question the form of language itself: "and is it a or the / or is it singular plural? / who are you to say / how s should be placed / if? [...] who are you to say / a or the?" (weiss 1977, n.p. ["Second Day"]). The empty and yet ever-changing desert "that by its very barren-ness / holds all possibilities" (weiss 1977, n.p. ["Fifth Day"]), becomes the perfect scenario for the poet to expand the poetic vision beyond the different mechanisms of control that, as shown in the previous section, affect both genders. Though this is definitely not easy – "it is a hard thing / for an animal / to break habit" (weiss 1977, n.p., ["Ninth Day"]) – the speaker in *Desert Journal* urges the reader to refuse the bounded limits of oneself or one's body to become "fluid and evolving, uncontainable and unquantifiable" (Carden 2015, 88). It is in the recognition and elevation of the complexities and contradictions of the "multi-singular" (weiss 1977, n.p. ["Twenty-first Day"), that the speaker in *Desert Journal* manages to use language beyond social, aesthetic, and linguistic constrains. In weiss' poetry, then, language is understood in its complexity, used with caution as it can participate in both the perpetuation, as well as the contestation or elimination of the categories men and women are placed into.

The ability to break and bend language to her will comes naturally to German-born weiss, who has expressed her preference for English as a "new + crude language [...whose] rules are riddled with exceptions – unlike other tongues that make for a rigid tradition," as she wrote in the introductory speech for a class she taught at the Schule für Dichtung (Vienna Poetry School) in Vienna (1998). It is in this context of language unruliness that she contests the differences between men and women by literarily and literally dismembering the linguistic symbols that contain them – see, for instance, these lines from "altarpiece": "we are all human / who man / wo man/ we are / all / YES" (2012b, 42). Anne Waldman shares weiss' conception of language as a double-edge sword. For instance in "Sprechstimme (Countess of Dia)" in *The Iovis Trilogy* (2011) Waldman denounces the use of different systems of oppression that erase voices of lives that become eternally suppressed by a power structure that binds them to silence or death: "=how many plagued (you knew this was no lie)= / =or censored (slow death)= / =possibly strung out=" (623). Despite these dangers, and the domineering discourses that deny the experience of different minorities, Waldman chooses to elevate poetry as a tool to contest so-called realities, a place to reinvent not only oneself but also language: "I will re=-in=vent my roles / [...] / sleep in the margins of my writing / speak there too" (627). Speaking from the blurred edges of the inner desert, much like the margins of Waldman's epic poem, the speaker in *Desert Journal* uses language to liberate the self and the artistic creation: "without a plot / the story has a chance / to make it / on its own" (n.p., ["Twenty Fifth Day"]).

The reinvention or reconfiguration of a language intimately linked to the dissolution of the poetic "I" lies at the heart of weiss' expansion of poetry[6] in "Light" (1976), a series of poems written in response to the light shows of the visual artist Elias Romero. Like the poem "Africa" in its sort of received composition[7], the poems in "Light" seem to have a life of their own, follow their own logic and rules. Indeed, the poetic "I" in this series, as if fusing with the darkness of the room and the fast-changing shapes and images created by the paint and oils, is continually shifting, multiplying, appearing and disappearing. Like the broken syntax and splintered images – "blister blister yellow / is spread to naked next to / when the inundation is" (27) – the speaker is introduced as "intangible" (27) and multiple:

> all my selves are returning
> to tell me the story of travels
>
> one caught the green dream
> in a still-song
> and waiting
> another rivered
> the two close planets of one
> one shattered the dream (27)

As if to keep up with the speed of the lightshow, the speaker subdivides into different selves, all impersonal and de-genderized, that allow her to transpose Romero's art into her own artistic medium. It is not until later on in the series that an "i" and a "you" enter the poem. Nevertheless, these are two very volatile presences which appear, just as the colors and shapes of the light show, in constant state of flux. This freedom, as it happened in *Desert Journal*, necessitates a break from the past to free the "I" from itself – "i broke myself / when i gathered the past / i left me / scrambling eggs for help" (32). Passing moments of recognition and self-identification vanish within a few lines; for instance, in "Five C", the speaker's assertive "here i am" (48) when she sees herself reflected "in an orien-

---

[6] For an analysis of weiss' work in relation to the concept of Expanded Poetry, see Encarnacion-Pinedo's "Expanded Poetry and the Beat Generation: The Case of ruth weiss" in Nancy M. Grace's *The Beats: A Teaching Companion* (forthcoming).

[7] In "I Always Thought You Black" she recounts the moment she read the poem to the South African manager of a dance troupe that was performing in San Francisco: "i'm in the poem now. the words sing out. is this my voice. [...] but that is the story of my village. how do you know this. it was given to me." (2011, 45)

tal / moon-face" (48), quickly disappears as the fluid paint dissolves the moon: "nothing will shatter / that which is / though i dissolve with it" (48). Once again, these images are loosely linked to experience lived through the double role of speaker of the poem and the spectator of the show, as they appear to her "trembling to be recognized" (44). Nevertheless, even if the images seem to be directed to a singular and specific self – "the bird squat / one tear & quiver all resolve / to frenzy on my eye world" (50) – this "I" is far from a stable entity, but rather changes and dissolves as fast as the images themselves. In "Six a", for instance, the speaker is "a 1000 thing," and in "Seven a," the images lead to "a final none eye" (54). The eye/I performing the double function of the spectator and the poet finally dissolves at the end of the collection. In "Seven b," in the context of a continuous flow of time – "the now is always past / the now is always once again" (56) – the speaker, and with her, her interpretation, disintegrate: "i am melting melting / no more me / no more / nulla-la" (56).

Allowing the dissolution of identity and gender in her poetic "I" in these examples, weiss foregrounds the autonomy of the artistic creation. No longer affected by the limited and limiting discourses that often created the conflict between the genders in the poems analyzed in the previous section, the poetic "I" is now permitted to carry out its vision without restrictions; the desert's ever shifting "grain to grain / all colors" (1977, n.p. ["Twenty-second Day"]) allows the poetic "I" to dwell in "words beyond meaning / sounds beyond words / silence beyond sound" (1977, n.p. ["Twenty-seventh Day"]) in an approach that very much mimics the liberating creative space of the light show, which is also "beyond border / beyond color" (1976, 54). Nancy Grace has aptly noted how *Desert Journal* "rejects tribalism, confession, and identity politics for the allure of prophetic and transcendental artistic individualism." (60) While this is true for most of weiss' body of work – even in seemingly more personal poems such as "Single out" or "Full Circle" – in *Desert Journal* and "Light", as well as other works in which she uses poetry to "enter" other mediums and spaces[8], the dissolution of a gendered and domineering poetic "I" becomes another strategy through which the poet frees the poetic vision itself from fixed categories, whether based on gender, genre or the medium specificity art happens through. This links weiss' work with postmodern aesthetic practices such as fragmentation and minimalism, an issue which Ronna Johnson, although not referring to weiss, sees as the "Beat Movement revolution [...] the shift, at its best and to rare-

---

8 Examples include the first section in *South Pacific* (1959) as well as the unpublished "i enter the painting of Marianne Hahn", a series of poems written in response to the work of the abstract and portrait painter Marianne Hahn.

ly realized, to gender transcendence". (2017, 176) Favoring spontaneity, synchronicity, and serendipity throughout her work, weiss' fluid voice hybridizes the poetic verse at the same time that creates an alternative route from which to transcend the social and aesthetic spaces in which the poet participates; this space, which constantly questions "How real is i?" (1977, n.p. ["Thirty-fifth Day"]), resists ready-made categorizations.

Conclusion: poet as instrument

In the works analyzed in this chapter, and in her overall body of work, weiss seems particularly aware of the dangers of writing *as*, *for* or even *about* man or woman as fixed categories. Having exposed the dangers of such discourses, weiss favors an approach focused on lived experience above and beyond exclusionist or reductionist divisions between the sexes. Much like Gayle Rubin's predicament in the influential "The Traffic in Women" (1975), in which she writes that "we are not only oppressed as women, we are oppressed by having to be women, or men as the case may be" (204) – weiss frees both women and men from the social and artistic strictures that hold them down. When asked to curate a women's reading series in San Francisco in the 1970s, as Thomas Antonic relates (2015, 181), weiss' condition to accept the invitation was that men would need to participate as well, as they too, just like women, could be unjustly ignored by literary circles. Not a mere utopian stance, nor an act of obliviousness towards the actual position women and women artists occupied and occupy[9], the ultimate goal of weiss' egalitarian and inclusive poetic "I" is to let the poetic vision – whatever medium it may be expressed through – run free and unaffected by external factors that include, as we have seen, the very identity of the poet.

Ultimately, weiss' de-genderizing and expansive poetics highlight the poet's insistence on eliminating the barriers, obstacles and divisions that separate and damage women and men. The expansive consciousness that allows the poet to write *beyond gender* – as she does in *Desert Journal* and "Light" – complicates the arbitrary literary labels based on sex or gender. As a case in point, using the elasticity of gender to extend labels, weiss uses poetry to expand and contest the category "Beat woman" by including the hermaphrodite Elsie John in her "For these Women of the Beat". Brought together through "the beat the beat the beat" that introduces and glues the women in the collection together,

---

[9] She was quite aware of this: Ferlinghetti refused to publish her early work on the grounds that they did not publish women. (Antonic, 182)

weiss acknowledges, as this chapter has shown, that in poetry and art, "it's all carnival / either sex or both" (2012a, 22). Either sex, both, or none, "goddess of the Beat Generation" or not, weiss remains a poet of her own.

Works Cited

Antonic, Thomas. "From the margin of the margin to the 'Goddess of the Beat Generation': ruth weiss in the Beat Field, or: 'it's called marketing, baby." *Out of the Shadows: Beat Women Are Not Beaten Women*, edited by Frida Forsgren and Michael J. Prince, Agder: Portal, 2015, pp. 179–199.

Borcich, Zida. "Poetry Matters: ruth weiss." *Real Estate Magazine*, vol. 27, no. 1, 2013, pp. 2 +. Print.

Breines, Wini. "The Other Fifties: Beats and Bad Girls." *Not June Cleaver: Women and Gender in Postwar America 1945–1960*, edited by Joanne Meyerowitz, Philadelphia: Temple UP, 1994, pp. 382–408.

Carden, Mary Paniccia. *Women Writers of the Beat Era: Autobiography and Intertextuality*. Charlottesville and London: University of Virginia Press, 2018.

Daly, Mary. *Gyn/ecology: The Metaethics of Radical Feminism*. Boston: Beacon Press, 1990.

Damon, Maria. "Victors of Catastrophe: Beat Occlusions." *Beat Culture and the New America: 1950–1965*, edited by Lisa Phillips, New York: Flammarion, 1995, pp. 141–152.

Davidson, Michael. *Guys like Us: Citing Masculinity in Cold War Poetics*. Chicago: University of Chicago Press, 2004.

di Prima, Diane. *Loba*. New York: Penguin Books, 1998.

di Prima, Diane. *Recollections of my Life as a Woman: The New York Years*. New York: Penguin Books, 2001.

Ginsberg, Allen. *Selected Poems: 1947–1995*. London: Penguin Books, 1996.

Goodman, Paul. *Growing up Absurd*. 1960. New York: New York Reviews Books, 2012.

Grace, Nancy M. "ruth weiss's *Desert Journal:* A Modern-Beat-Pomo Performance." *Reconstructing the* Beats, edited by Jennie Skerl, New York: Palgrave Macmillan, 2004, pp. 57–72.

Grace, Nancy M., and Ronna C. Johnson, eds. *Breaking the Rule of Cool: Interviewing and Reading Women Beat Writers*. Jackson: UP of Mississippi, 2004.

Johnson, Joyce. "Beat Queens: Women in Flux." *The Rolling Stone Book of the Beats: The Generation and American Culture*, edited by Holly George-Warren, New York: Hyperion, 1999, pp. 40–49.

Johnson, Ronna C. "The Beats and Gender." *The Cambridge Companion to the Beats*, edited by Steven Belleto, Cambridge: Cambridge UP, 2017, pp. 162–178.

Kaufman, Bob. "Bagel Shop Jazz." *Collected Poems of Bob Kaufman*, edited by Neeli Cherkovski, Raymond Foye, and Tate Swindell, San Francisco: City Lights, 2019, pp. 11–12.

Mailer, Norman. "The White Negro." *The Portable Beat Reader*, edited by Ann Charters, New York: Penguin, 1992, pp. 582–605.

Rubin, Gayle. "The Traffic in Women: Notes on the 'Political Economy' of Sex." *Toward an Anthropology of Women*, edited by Rayna R. Reiter, New York: Monthly Review Press, 1975, pp. 157–210.

Waldman, Anne. *Kill or Cure*. New York: Penguin Books, 1994.

Waldman, Anne. *The Iovis Trilogy: Colors in the Mechanism of Concealment*. Minneapolis: Coffee House Press, 2011.

weiss, ruth. *Steps*. San Francisco: Ellis Press, 1958.

weiss, ruth. *South Pacific*. San Francisco: Adler Press, 1959.

weiss, ruth. *Blue in Green*. San Francisco: Adler Press, 1960.

weiss, ruth. *Light and Other Poems*. San Francisco: Piece & Pieces Foundation, 1976.

weiss, ruth. *Desert Journal*. Boston: Good Gay Poets, 1977.

weiss, ruth. *Single Out*. Mill Valley: D'Aurora Press, 1978.

weiss, ruth. "Intro Speech." ruth weiss Papers. BANC MSS 99/84. Bancroft Library, University of California, Berkely, CA, 1998. Unpublished manuscript.

weiss, ruth. *Full Circle / Ein Kreis Vollendet sich*. Bilingual Engl. / Germ. German transl. Christian Loidl. Vienna: Edition Exil, 2002.

weiss, ruth. *No Dancing Aloud: Plays and Poetry / Lautes Tanzen nicht erlaubt: Stücke und Gedichte*. Bilingual Engl. / Germ. German transl. Horst Spandler. Vienna: Edition Exil, 2006.

weiss, ruth. *Can't Stop the Beat: The Life and Words of a Beat Poet*. Studio City [Los Angeles]: Divine Arts, 2011.

weiss, ruth. *A Parallel Planet of People and Places: Stories and Poems*. Zirl: Edition BAES, 2012a.

weiss, ruth. *A Fool's Journey: Poems and Stories / Die Reise des Narren: Gedichte und Erzählungen*. Bilingual Engl. / Germ. German transl. Peter Ahorner and Eva Auterieth. Vienna: Edition Exil, 2012b.

Whaley, Preston Jr. *Blows like a Horn: Beat Writing, Jazz, Style, and Markets in the Transformation of U.S. Culture*. Cambridge: Harvard UP, 2004.

Image Source

ruth weiss in 1973, photographed by Scott Runyon. Courtesy of ruth weiss.

Polina Mackay
ruth weiss and the Poetics of the Desert

> *Through a desert she hurried,*
> *thirsting she ran*
> *to reach becoming,*
> *passed three water holes*
> *but never saw them,*
> *so eager was she to reach*
> *outward evidence*
> *of her inward drawing*
> (Madeline Gleason, "Once and Upon," 126)

A. Robert Lee writes of *Desert Journal* (1977) that the book is ruth weiss' "mapping of Beat-like poetic mind and spirituality" (2). The book is indeed relevant to the Beats but also to a wider discourse of spirituality and poetics. *Desert Journal* offers a journey of spiritual self-discovery. As weiss herself notes of the book in an interview with Thomas Antonic, "And when you read it [...] open it in your heart. Don't try to get a meaning or analyze. You've got to enter it from your own center" (qtd. in Antonic, 194). However, this process of unearthing is filled with ambiguity, first made apparent in the ambivalent title of the book. Does the title describe a diary in the sense of a record of one's daily activities? Or does it refer to a journal, a record of one's feelings, thoughts and meditations on their own experience within a wider historical, cultural or even sociopolitical context? In more ways than one, the book is both. On the one hand, it charts the speaker's daily routines, over a period of forty days, going as far as to number each day. On the other hand, the book elaborates on these routines and links them to extended meditations on subjects such as religion and the arts. What is more, the pieces of writing in *Desert Journal* are clearly poems, evident from the form, repetitions of sound and the compressed language in the word choices. It also dawns on the reader very early on that the poems are accentuated by two significant theological narratives of Judaism and Christianity, namely Moses' forty years and Jesus' forty days and nights in the desert. But the poems also go beyond a simplistic retelling of these stories as they depict the desert as a significant symbol in its own right. The poems write the desert as a symbol of both hardship and hope, of isolation and togetherness, or of death and life. They feed into a wider tradition of poetry which treats the desert as a multifaceted symbol: examples include British Romanticism (poems by William Wordsworth and Percy Shelly); modernism (such as Robert Frost and T.S. Eliot); and contemporary work specific to the desert in the American imagination (e.g., the poems about the Iraqi desert by such poets as Sandra Osborne). We will return to these intersections later on. Although the book's images of the desert build on the landscape from the Judaic and Christian stories and on the existing poetic tradition of desert imagery, they are also distinctly American. By the end of the book, the desert is reconceptualized as also the ultimate symbol of America in recurrent images that capture the fluidity of the landscape with focus on the sand or the changing light on the sand dunes. My chapter will argue that, taken together, these three significant layers of weiss' book – the Judaic and Christian narratives, the desert as a poetic symbol and the desert as embodying America – amount to a poetics of the desert in weiss' work.

The Religious Context of *Desert Journal*

First, it is important to note to which extent the religious context comes into play in *Desert Journal*. It is obvious from the beginning that the book reflects the journeys of Moses and Jesus through the desert. This mirroring is obvious not just in the highly symbolic use of the number 40 (40 poems for 40 days), but also in the use of the desert as the book's setting. This theological context is visible right from the start, even in the prologue-poem that precedes the poems in the main section:

> you are entering a certain desert
>
> like stones or bones
> marking sand
>
> flame & cloud
> things with wings
> call your number
> read your day
> see if it talks to you alone
> like stone or bone
> in sand
>
> other days
> other ways—(n.p.)

This piece is a blueprint on how to read *Desert Journal*, offering advice on how to appreciate each poem ("read your day / see if it talks to you alone"). At the same time, the piece is a poem itself made so by its use of poetic devices, including figures of speech and repetitions of sound. Note the simile at the start ("you are entering a certain desert / like stones or bones / marking sand"), the rhyming sounds in "alone" and "bone" and in "days" and "ways," and the developing image of the desert as characterized by marks in the sand, "flame & cloud" and "things with wings." At the same time, the image recalls the story of Moses who led the Israelites out of Egypt with the help of God, guiding them by cloud during the day and by flame during the night, as it is recorded in *Exodus* 13:21 ("And the LORD went before them by day in a pillar of a cloud, to lead them the way; and by night in a pillar of fire, to give them light; to go by day and night"). weiss develops the images of fire and cloud as guiding lights further in the poems that follow this introduction. In particular, poem "Thirty-Third Day" describes a developing image of people gathering around a small flame as a ritual of community building. The poem is also set against the backdrop of persistent rain in the desert. Here the rain feeds into the poem's pathetic fallacy where the speaker's inner turmoil is reflected in the harsh weather conditions of the desert:

> when it rains in the desert
> it rains
> limbo-gray the day
> the rain keeps pouring
>
> the only warmth is fever
> galloping relentless hoovers of tears
> that leave one tearless
> rocked with dry dry laughter.

The poem also recalls God's punishment of the Egyptians' for their enslavement of the Israelites. When God decides to help the Israelites escape Egypt, he instructs Moses to "stretch forth [his] hand toward heaven" (*Exodus* 9:22); "And Moses stretched forth his rod toward heaven: and the LORD sent thunder and hail, and the fire ran along upon the ground; and the LORD rained hail upon the land of Egypt" (*Exodus* 9:23). This image closely resembles the combined imagery of fire and rain of *Desert Journal*, but in weiss the biblical story is framed by the person in the poem. In "Thirty-Third Day," for instance, the speaker inundates the theological narrative with images of individual, personal happiness and suffering due to sexual love. The poem speaks of a lover who "appears / to disappear again & again," causing pain ("the lover who must reach you / in such pain"), but also representing hope evident in the lover's connection to light:

> the lover who is now
> the moment that all the others paled
> a light so bright
> that one small flame
> could never be hover
> but stand up straight in the rain

> and all the rain
> could never douse that flame.

Here weiss fuses the story of God's wrath against the Egyptians with a personal narrative told in a confessional style. The poet, then, finds the personal in the biblical context while, conversely, the story of Moses and the Israelites is given a current feel in the machinations of modern love.

It is in the poems that elaborate on the image of the cloud that weiss makes the best use of the theological context in *Desert Journal*. This makes sense as the cloud in *Exodus* has great symbolic significance: the glory of God appears to the Israelites in the cloud (16:10); the Lord comes to Moses in a thick cloud (19:9); when Moses goes up the mountain, God calls him "out of the midst of the cloud" (24:16); Moses goes into the midst of the cloud where he stays for 40 days and 40 nights (24:18); and finally God descends in the cloud (34:5), becoming "the cloud of the LORD" (40:38), the guiding light that the Israelites were to take up to continue on with their journey. In essence, the cloud represents the manifestation of God to Moses with the forty days and forty nights Moses spent in the midst of the cloud as a unique rupture in time where humankind gets the opportunity to have a brief encounter with the divine. weiss uses the image of the cloud as a central feature of the poem "Twelfth Day." Here the cloud appears at night to brighten the desert, but also becomes a manifestation of the loneliness and sense of alienation that the landscape of the desert can invoke. The night under the light of the cloud in the desert is described here as "solemn," an allusion to the language of religious texts bringing to mind sincerely meant declarations of "solemn oaths" to God.

As in many religious texts, in *Desert Journal* the story of humankind and its dealings with the divine is determined by the narrative of sin. It is sin and the need for atonement that drive Moses to Mount Sinai. weiss addresses sin head on in the penultimate poem of *Desert Journal*, where she urges the reader to "LET SIN GO" and

> no more must
> the rusty hinge of the past
> unhinged
>
> falling
> falling
> free
>
> […]
>
> LET GO SIN

> all of me
> part of me
> none of me

weiss' idea of freedom transpires as emerging at a moment where the bearer of sin lets go and sin lets go of the bearer in the same instant. While this image may be reminiscent of Moses and the Israelites' search for redemption, it is also a direct writing against the narratives of the past. Specifically, it is an act of writing against a significant part of the biblical story of Moses in the desert, the part in which God passes the ten commandments against sin to Moses. This is evident in weiss' description of the freeing from the past as predicated on the demand for "no more must." At the same time, this freeing is also described as "unhinged," which means deranged. The use of this specific word suggests that weiss likens the freedom acquired from breaking away from past narratives to a form of delirium, but surprisingly rather than causing confusion in the speaker this new state of mind offers clarity of vision.

This lucidity is expressed clearly in the theological layer of the last poem in the collection in which the poet announces that "the desert become[s] clear," while the speaker herself is "lighter than light." Having linked light to God consistently earlier in the book in the images of fire and cloud in the appropriation of the story of Moses in the desert, weiss writes the speaker's discovery of their vision and voice (which is now recognizably female having been androgynous or transgender in the earlier poems) as a form of religious-like epiphany, where the poet has a revelation following the manifestation of the divine.

The debt to the biblical story of Moses in weiss' work helps us determine the emergence of the poet's vision towards the end of *Desert Journal* as primarily spiritual. For the story of Moses and the Israelites wandering through the desert is also a story about the search of identity. The story of Moses and the Israelites in *Exodus* stresses the need for a form of collective identity expressed in the sense of community by the Israelites and in God's desire to treat them as a people. *Desert Journal* repeats the sense of community as an important part of being human – namely in the use of the generic pronouns "you," "one" and "we," in the recurrent references to community rituals (such as singing or dancing), or in the images of people journeying like herds (e.g., in poem "Ninth Day"). An equally important figure in the text is that of the lonesome traveler. This aspect of the journey in the desert is crystallized when we take into account the second major theological narrative that weiss uses in *Desert Journal*, which is the biblical story of Christ wandering through the desert for forty days and forty nights. This story, known as "The Temptation of Christ," narrates Jesus' wandering through the Judaean desert after he was baptized by John the Baptist. In the desert, Christ is tempted by the devil, but he manages to resist all of Satan's temptations before returning to Galilee to take up his ministry. *Desert Journal* retells this story in a central image of a lone figure traveling through the desert's harsh terrain. weiss chooses to label this figure "the wanderer," who alone has to face the elements of the desert. Often the "desert-wanderer" keeps going but "moves even more slowly" when they have to face harsh weather conditions, like the wind in the desert. This image creates a laid-back atmosphere in the book, which is counteracted by the recurrent images of threatening birds flying through the desert as if stalking their prey on the ground. The penultimate poem of the collection builds up the image of the lone desert-wanderer and the ominous birds hovering above to an allegory for Christ's escape from temptation and, by extension, the poet's own freeing from the past:

> wings
> dark things hover
> cover the desert
> as far as the eye cannot see
> waiting waiting
> for an open wound
>
> wind is a higher game
> [...]
> and the wanderer gazes dazed
> at his open wounds closing
>
> wings dark things change shape
> move not up
> but away
> [...]
> the worst in the desert
> is the past
>
> come pit
> come pendulum
> do me in!
>
> LET GO SIN

Here weiss makes use of the symbol of the wound in Jesus' story, namely the image of the open holy wounds suffered by Jesus during his crucifixion and the idea of Christ being able to heal the wounds of others. In addition, weiss brings human suffering to the forefront through an allusion to Edgar Allan Poe's famous short story "The Pit and the Pendulum" (1842) in which a prisoner of the Spanish Inquisition is tormented. Here weiss equates human suffering with the pain of Christ, while Jesus' tormenting in the desert by the devil is given a human dimension. These two layers act as frameworks that tell us how to read the freeing of the poetic voice in the collection's penultimate poem. As weiss brings the godly and the human together, the release of the poetic voice is both a spiritual release (God) and a physical liberation (human). Essentially, weiss points to the freeing of the mind through writing but also to the literal act of speaking in performance. It is worth remembering here that *Desert Journal* is both a printed book but also a performance piece, which weiss performed countless times over many years. Christ, as the symbol of the coming together of the divine and the human and the spirit and the flesh, embodies this duality. weiss' appropriation of the biblical stories of Moses and Jesus suggest that the desert in the book is first and foremost a theological symbol. As such, it calls for a theological reflection on the meaning of experience and spirituality – as Moses and Jesus had to do in their own journeys through the desert. It further brings to the surface the idea of pilgrimage, a central concept in many religions, whereby the journey is a search for not just spiritual but also moral significance.

The Desert as a Poetic Symbol

The theological narratives and their emphasis on spiritual self-discovery act as a framework for weiss' main thesis in *Desert Journal*, which is that a collection of poems, whether read or performed, is inevitably a journey into the interior space of the mind. This proposition is abundantly clear in the prologue-poem of the book in which the speaker talks of entering a new space, where individual poems gain subjective significance for each reader. While the importance of the mind is also apparent in the theological narratives weiss uses, the mind as a complex concept becomes even more relevant to *Desert Journal* when we see it as a gateway into poetic imagination. Considering the desert in weiss as a poetic symbol relevant to the machinations of a poet's imagination opens up meanings further. For in this schema *Desert Journal* can also be read as a mapping of how a poet's imagination works.

First, it is important to note that there is a long tradition of poetry in English that treats the desert as a symbol. One of the first poets in modern times to use the desert as a potential source of symbolic significance is William Wordsworth. In Book 5 of *The Prelude* (1805), for example, the poet records a friend's dream of the desert,

> Sleep seized him and he passed into a dream.
> He saw before him an Arabian waste,
> A desert, and he fancied that himself

> Was sitting there in the wide wilderness
> Alone upon the sands. Distress of mind
> Was growing in him when, behold, at once
> To his great joy a man was at his side,
> Upon a dromedary mounted high.
> He seemed an arab of the Bedouin tribes;
> A lance he bore, and underneath one arm
> A stone, and in the opposite hand a shell
> Of a surpassing brightness. Much rejoiced
> The dreaming man that he should have a guide
> To lead him through the desert. (lines 70 – 83)

The theological layer of this extract is quite obvious with the desert understood as the locale where a guiding light towards spiritual growth may be found, as in the biblical stories of Moses and Jesus. At the same time, Wordsworth's dreamer romanticizes the desert as simultaneously a place of fascination and danger, inhabited by symbols that the poet is called upon to decipher – the Arab of the Bedouin tribes, the lance, the stone, the shell, the brightness as well as the distress and joy felt by the dreamer. In a short space of time, the poet writes the desert as a symbol of both war and peace, darkness and brightness, anguish and pleasure. A similar conceptualization of the desert occurs in Percy Shelley's poem "Ozymandias" (1818), in which the poet reports an encounter with a traveler who talks of the ruined statue of Ozymandias (an alternative name for the Egyptian Pharaoh Ramses II) in the desert, its ruin somewhat fitting of the potential destructive power of the harsh climate of the desert. For Shelley's traveler, the desert affords the statue symbolic significance, the lifelessness of the shattered stone complementing "The lone and level sands" (395) of the desert.

The British Romantics treated nature as occupying a pivotal place in the process of self-discovery, which they saw as an important step towards poetic completion. It is no surprise, then, that their depiction of the desert feeds into this wider project of self-realization through a poetic appreciation and re-imagining of nature. Wordsworth's and Shelley's description of the desert also illuminates the Romantics' focus on the continuous interplay of the external world with the poet's internal sphere. They express this simultaneous externalization and internalization by having a separate voice speak to the poet (a dreamer in the example from Wordsworth and a traveler in the example from Shelley). The effect in both cases is to present the desert as the space where the external and the internal worlds meet, where the desert is both an exterior landscape as well as an inner state of mind. Another poet, who has also been heavily associated with nature poetry, Stephen Crane similarly internalizes the desert as part of the speaker's psyche. In his desert poems, "I Walked in A Desert" (1895) and "In the Desert" (1895), Crane presents the space as a symbol of the multifaceted nature of human psychology. Both poems repeat the Romantic poetics of dialogue – in Crane's case with creatures that are not wholly separate from the speaker. In "I Walked in A Desert," the other is merely a disembodied voice that informs the poet that there is no desert, despite the speaker's experience of walking into the desert and experiencing "The sand, the heat, the vacant horizon" (v). Thus, the desert is realized as a state of mind. In the poem "In the Desert," the voice is that of a bestial creature, who eats its own heart in the desert (12). Again, the effect here is to internalize experience and imagine the desert as part of the speaker's solipsistic universe.

This internalization process of the external environment, whereby the desert becomes a principal symbol of isolation, is not confined to nature poets of the nineteenth century. In twentieth-century poetry, the desert clearly appears as an internal landscape of the mind. An excellent example of this is Robert Frost's "Desert Places" (1933) which compares nature in winter – the darkness, the cold, the hibernating animals and the empty spaces – to the loneliness the speaker feels in his heart, concluding that the winter outside does not scare him as he has his own "desert places" (183) within him. The idea of the desert becomes here a metaphor for the alienation felt when winter takes hold, and for the sense of abandonment humans feel when faced with isolation. A similar conflation of the external environment with the poet's emotions is repeated in Josephine Miles' poem of near the same time entitled "Desert" (1934). Here the poet talks of a natural wisdom gained in the desert, which enables her to feel the drought, the dryness and the heat and to know instinctively what to touch and what not to touch in the desert. In Miles' poem, the speaker becomes one with the desert in a conflation where the desert and its natural elements – the dry air, the cacti or the scorched stones – become manifestations of the speaker's emotions:

> When with the skin you do acknowledge drought,
> The dry in the voice, the lightness of feet, the fine
> Flake of the heat at every level line;
>
> When with the hand you learn to touch without
> Surprise the spine for the leaf, the prickled petal,
> The stone scorched in the shine, and the wood brittle;
>
> Then where the pipe drips and the fronds sprout
> And the foot-square forest of clover blooms in sand,
> You will lean and watch, but never touch with your hand. (15)

At the center of this poem, as in the British Romantic poetry of Wordsworth and Shelley mentioned earlier, there is a deep appreciation for the desert, which becomes a symbol of life. This is evident in the various images of nature's regeneration, such as the sprouting of the fronds and clover blooms, which grow even against the sand and the heat. This is once again a clear romanticization of the desert, an idea which is found in many other poets of the twentieth century. Examples include Harriet Monroe's "A Garden in the Desert" (1914), a travel poem that describes the colors, feel and smell of flowers in the desert in a dream-like state, noting the play of light and the slowing down of time against the sun: "So light and soft the days fall – / Like petals one by one / Down from yon tree whose flowers fall / Must vanish in the sun" (227). Mina Loy's "Mexican Desert" (1921) also depicts the desert as having an ethereal quality by focusing on the interplay between light and shadow in the landscape's various images, such as the image of the "hunch-back palm trees" (672) against the twilight. A more recent example of a poem imagining the desert as part of a dream is Yasmin Khan's "The Desert Night" (2013), which mirrors the fluidity of the desert's landscapes in the images of sand falling through the speaker's fingers, waves slumbering against the sand and the light from the moon becoming liquid silver running through the poet's body and soul ("Cup my hands to hold the liquid silver / To soothe my soul burnt under the sun").

Collectively, the examples above show how modern and recent poetry follows in the footsteps of the British Romantics in imagining the desert as having a dream-like quality. Thus they concentrate on the fluidity of the desert landscapes by focusing on images of the sand, the interplay of light and shadow, or the instances which affirm both death and regeneration. The effect is to turn the desert into a multifaceted symbol, which feeds into a wider appreciation of nature and its various elements. It has to be noted, however, that there are numerous examples in modern and recent poetry that depict the desert and its harsh living conditions for humans as a metaphor for the decline of humanity and civilization. Unsurprisingly, many of the great examples of this kind of poetry come from work written between the two World Wars or just after the end of the Second World War, a time when the potential loss of the world's sense of humanity was most apparent in the war atrocities globally and in the Holocaust (Sherry 2003; MacKay 2009). The most famous example of this kind of poetry is the work of T.S. Eliot, specifically *The Wasteland* (1922) and *Four Quartets* (1943), which are filled with images of waste and degeneration. However, Eliot's books are also reflections of humanity's attempts to find meaning amongst the ruins of war. *The Waste Land*, for instance, describes this attempt as, David Moody has successfully argued, "fill[ing] all the desert with inviolable voice" (115).

A more bleak vision of the desert as an apt locale for war and devastation is found in the recent poems by American writers about the Iraqi desert. Inspired by the war waged against Sadam Hussein's Iraq by the US and allies, these poets see the Iraqi desert and its unfamiliar landscape (to Americans, of course) as embodiment of the futility of a war conducted very far from American borders. As we read in Sandra Osborne's "Iraqi Desert" (2004), the war is primarily a mediated experience that, nevertheless, still captures the bleakness and loneliness of the desert: "So flash / The sandy TV screens / So bleak and all alone". Osborne's poem feeds into an emerging tradition of contemporary American poetry that contemplates America's twenty-first-century wars. A good example is Seth Brady Tucker's "The Road to Baghdad" (2011) which does not mention the desert, but is similarly filled with images of deadly silence and burning heat (online). Elyse Fenton's Baghdad poem "Word from the Front" (2011) goes further than the desolate spaces found in other poems of this type to note the varied landscape of Iraq, as it is mediated to the speaker through her lover who is an army pilot and presumably involved in direct attacks against the city. Her poem is accentuated with the landscape variation as it might be seen from a pilot's perspective; it notes "the verdant green / neck of Baghdad's bottle-glass night" and the "cratered ground". A common thread in these poems is a sense of isolation felt by the speakers, a group which includes soldiers as well as loved ones who are at home.

As in the tradition of desert poetry I have outlined above, weiss' *Desert Journal* treats the desert as a complex symbol adding layers to an already highly symbolic narrative that, as shown earlier, is accentuated by references to the biblical journeys of Moses and Jesus. What is more, like many of the poets who use the desert as a central metaphor in their works, weiss also writes this journey as a process of self-discovery centered on the idea of transformation. The central image of the very first poem of *Desert Journal* is of a flying bird that changes into a kangaroo bird of either sex or both that lands in the desert to begin its journey. Almost every poem that follows this opening is filled with images intent on giving the book an overall dream-like quality where all kinds of transformations are possible. Poem "Sixth Day," for instance, depicts the desert as a place of mirage and illusions:

> the desert is fine
> it conjures all places
> this is considered mirage
> ageless
> without mirror
> it is considered illusion—
> that pure-point where all gathers

> one dreams of the tree
> another of the sea
> one dreams of the root
> another of the foot
> one dreams of the hand
> another of the sand
> that makes its thought
> mountain upon mountain

As Wordsworth's dreamer, the dreamers in weiss' poem transform the desert landscape into a series of personal visions that give a snapshot of their inner worlds. The terrain of the desert becomes one with the psyche of the person in the poem, in a process where the gazer/speaker/poet shapes the external world with their mind as they, in turn, are changed by the surrounding environment. In this respect, *Desert Journal* seems to be an extended version of Madeline Gleason's poem "Once and Upon," in which the speaker travels through space and time as they continually change. In the extract I used as an epigraph, the speaker hurries through the desert frantically to reach "becoming," self-realization. At the center of both Gleason's and weiss' vision of becoming is the idea of transcendence: the dreamers in the example above from *Desert Journal* seem to be having experiences that are beyond the normal or physical level as they seem to become one with their environment. Of course, the possibility of spiritual ascendancy is abundantly clear in weiss' decision to incorporate the biblical stories of Moses and Jesus, but it is worth noting further that she also layers all the other poems that do not overtly allude to the theological narratives with the idea of transcendence through poetry.

Transformation and ascendancy through poetry is mostly evident in the poems of *Desert Journal* in which the desert appears as a metaphor for the poet's inner world. A strong example is poem "Seventh Day," which compares the speaker's realization "that nothing is permanent" to waves slapping against each other. Once the poet creates this dream-like atmosphere of non-permanence, she offers a climactic moment of transcendent and exuberant epiphany in the desert: "AND ON THE SEVENTH DAY / THE DARKNESS SAW ITSELF!". The image and language here are reminiscent of the theological narrative about the birth of the world in *Genesis*, but it is the desert's reflection of the speaker's inner turmoil and joy that allow this moment to become a manifestation of the poet's psyche. The poem shows the desert reflecting one's own face, even in the midst of mirages and illusions, and is further compared to the human heart ("Twenty-First Day"). Towards the end of the book the speaker becomes one with the desert: "the city of oneself / crawls over the desert" ("Thirty-First Day"). As in the poems that capitalize on the desert's symbolic significance discussed earlier, weiss uses the connection between the desert and humanity to unlock a wide spectrum of emotions. Wordsworth talks of the simultaneous joy and sadness the idea of the desert brings; Shelley notes the hopelessness of humanity against time realized in the image of the ruined statue of Ozymandias in the desert; Miles, Monroe and Khan focus on the fluidity of the landscape to mirror the variety of nature and human experience. Similarly, weiss connects a range of human emotions to the experience of journeying through the desert, a few of which we have already explored. Among them are joy, fear and emotional pain. But in weiss this experience is both a spiritual and a physical experience with the soul rarely separated from the body. In poem "Twenty-Second Day," for example, weiss talks of physical and emotional pain as the first step into the desert with the suffering of soul searching compared to the burning sensation of the soles of the feet as they walk on the hot sand:

> all pain the burning sand
> the soul keeps moving
> as soles of feet
> on burning sand

This image of the torment of body and soul in the desert is an obvious allusion to the solitary journey of Jesus through the desert, where he experienced both spiritual and physical suffering. *Desert Journal* adds to the image of the tormented wandering Jesus through the desert the idea that, while the epiphany at the end of this journey is significant, the journey itself is just as important. More specifically, it is the act of moving (the body, the soul) that makes the journey worth having; it is the moving soul and body which make the epiphany happen.

The images above – the moving feet, the sand or the heat – all suggest that at the center of weiss' poetic vision of a journey through the desert is the idea of fluidity. This is abundantly clear in weiss' appropriation of the biblical stories of Moses and Jesus (after all they are both travelling stories), in the recurrent images of waves or sand and in the several narratives of transformation and transcendence in the book. Importantly, weiss layers the moments of transcendence with languages other than English, including German and Hebrew. A good example is the moment of epiphany in poem "Seventh Day," which is followed shortly after by a stanza written in German. weiss' insertion of languages other than English in the text brings to focus the poet's interest in how the fluidity of human experience may be expressed in the arbitrariness of language. Nancy Grace and Ronna Johnson rightly note that weiss' interest in linguistic play is illuminated in the poems' similarities with one of the author's favorite writers, Gertrude Stein (56). The poem on the fourth day, for instance, finishes with:

> back broken
> broken back
> the cat MONK
> allows his belly
> with cunt-markings
> to be stroked
> purring to death
> BE NOT AFRAID
> I AM READY
> FOR THAT SEPARATION
> THAT IS ONE

As Grace and Johnson point out, "These lines exemplify the way weiss wields doggerel, contemporary slang, neologisms, and word inversions to track the free associations of the moving mind" (56). weiss emphasizes the fluidity of language that *Desert Journal* contains in the performances of the poem where she would invite the audience to pick any number before reading the poem matching the day of the number (Grace 2007). This allowed spontaneity and unpredictability in the performance, making each show a unique experience. In addition, the performances underscored weiss' central metaphor in *Desert Journal*, which writes the desert as a symbol of the fluidity of language, poetry and human experience. As weiss put it in an interview, *Desert Journal* "is a perfect performance piece. It is living language with a definite focus and shifting patterns like sand in the desert" (Grace and Johnson, 67).

The danger of using the fluidity of the desert as a symbol for storytelling and poetic invention, going as far back as biblical stories, is that it has the potential to universalize human experience, to see life as defined by certain unavoidable happenings, such as spiritual epiphanies. weiss writes against the universalization of experience by first making her speaker androgynous or transgender to start with and female in the final poems of the collection[1]. The ambiguity created by the desert as a multifaceted symbol of, as we have seen, both light and shadow, life and death, resolve and uncertainty, is what makes gender indeterminacy in the work both possible and celebrated. Another way weiss writes against the idea of the desert as feeding into universal narratives of soul searching is by framing the space as distinctly American, using certain moments from American history as a lens through which the desert can be read. I explore this layer of *Desert Journal* next.

---

[1] For a recent analysis of the politics of gender in *Desert Journal*, see chapter 4 in Carden's *Women Writers of the Beat Era*.

The Americanness of ruth weiss' Desert

The uniqueness of weiss' *Desert Journal* is that, despite the equivocality inherent in poetic symbolism and in gender ambiguity, her depiction of the desert has a distinct character. weiss' desert is primarily American, dominated by American history, storytelling and landscapes. weiss depicts the desert as a piece of America, an idea made apparent in the poems' references to Native American languages and cultures with the speaker announcing in the first poem "i'm in Indian territory," while parts of the desert landscape are routinely labelled as indian, e. g., "indian sun" ("Third Day"). Towards the end of the book, the poet replaces indian territory with what she calls the "american way" of the "american word" and a "boston" attitude ("Thirty-Sixth Day"). This substitution alludes to the story of the colonization of the Americas by Europeans and the settler colonial narrative it created, which writes the essence of Americanness as possessing a particular way of life (the American Dream), the language of colonialization (English in the case of North America) and urban living (named here as the "boston" attitude). weiss' allusions to the settler narrative and the ensuing appropriation of Americanness by settler-Europeans suggest that the poet incorporates the recognition of this history into the process of spiritual and poetic self-discovery. Nevertheless, weiss continues throughout the book to interchange "american" and "indian," both words written in lowercase letters, as her own name, to deflate their connotative importance. But even then, the different parts of the desert landscape, which have become so important in the process of poetic discovery that we have identified earlier, are written as "indian." Thus, the book speaks of an "indian" sun looming over the desert ("Third Day"), while the indianness of the outer world is also internalized into the poetic mind:

> she feels her indian strong
> but the indian is not pure
> so she's not sure
> what is her right
> and what is her wrong. ("Twenty-Fifth Day")

Here the speaker rightly feels out of step with the native landscape of America, being as she is of white European origin and an integral part of the settler colonial mindset. The uncertainty weiss expresses here around the issue of the ownership of the American land ("so she's not sure / what is her right") determines her treatment of America in *Desert Journal* as a symbol with a fraught history around the issue of who owns the land that makes up America. What is more, weiss' nod to the idea of wrong or of a wrong being done in the same poem reframes the history of land ownership to a narrative of injustice. To put it simply, weiss alludes here to the injustice carried out by white settlers of European origin to the Native peoples of America. Instances like these make weiss' book a mapping of this particular chapter of American history.

weiss' perspective, as it comes across in *Desert Journal*, is Euro-centric and part of the colonial mindset. For incorporating the idea of the desert into the discourse of America fits well with twentieth-century European understandings of America. Two iconic European works on America, the travel book of philosophy entitled *America* (1988) by French philosopher Jean Baudrillard and the film *Paris, Texas* by German film-maker Wim Wenders (1984), depict the desert as the ultimate sign of America as a vast, unknowing and unknowable entity. Similarly, *Desert Journal* builds an image bank of desert landscapes organized around the idea of vastness. In poem "First Day," weiss talks of the "vast past" of the desert; in "Eighth Day," she depicts the desert as "vast & slow" realized as a fluid construct in its sand-filled open spaces. While at points the speaker despairs, calling the desert simply "too vast" and "too fast" "for human belief" ("Twenty-Sixth Day"), she then concludes that the problem is not the vastness in the end, but the past ("Thirty-Nine Day"). In other words, weiss returns to the fraught history of America, a discourse that by the end of the book emerges as fundamental to understanding *Desert Journal* fully.

Having said this, as we have seen, two other layers come into *Desert Journal* in a major way and have to be recognized as significant parts of the fabric of the text – the religious context through the Biblical stories of Moses and Jesus wandering through the desert and the literary history of the desert as a multifaceted poetic symbol. All three layers complement each other to provide *Desert Journal* with an elaborate narrative of the desert as the site of spiritual searching, poetic discovery or a journey into American history. In all these interconnected layers, weiss writes the desert as the manifestation of a series of never-ending imaginings. Ultimately, the desert for weiss captures the creativity, the need for searching, the inconsistency, the poetics, ambiguity as well as the limits and the simultaneous potential of the mind.

Works Cited

Antonic, Thomas. "From the Margin of the Margin to the 'Goddess of the Beat Generation': ruth weiss in the Beat Field, or: 'It's Called Marketing, Baby.'" *Out of the Shadows: Beat Women Are not Beaten Women*, edited by Frida Forsgren and Michael J. Prince, Kristiansand: Portal, 2015, pp. 179–199.

Baudrillard, Jean. *America*. London: Verso, 1988.

Carden, Paniccia Mary. *Women Writers of the Beat Era: Autobiography and Intertextuality*. Charlottesville & London: University of Virginia Press, 2018.
Crane, Stephen. "I Walked in A Desert." 1895. *Twentieth-century American Poetry*, edited by Dana Gioia, David Mason, and Meg Schoerke, New York: McGraw Hill, 2004, p. 5.
Crane, Stephen. "In the Desert." 1895. *Twentieth-century American Poetry*, edited by Dana Gioia, David Mason, and Meg Schoerke, New York: McGraw Hill, 2004, p. 12.
Eliot, T. S. "The Waste Land." 1922. *The Norton Anthology of English Literature: The Major Authors*, Vol. 2, 10th Edition, edited by Stephen Greenblatt & M. H. Abrams, New York & London: Norton, 2018, pp. 1326–1338.
Eliot, T. S. "Four Quartets." 1943. *The Norton Anthology of English Literature: The Major Authors*, Vol. 2, 10th Edition, edted by Stephen Greenblatt & M. H. Abrams, New York & London: Norton, 2018, pp. 1343–1349.
Fenton, Elyse. "Word from the Front." *Reed Magazine*, vol. 90, no. 2, 2011. Web: https://www.reed.edu/reed_magazine/june2011/articles/features/fenton.html. Accessed 26 Jan. 2021.
Frost, Robert. "Desert Places." 1933. *Adventures in Reading*, edited by Francis Xavier Connolly, San Diego: Harcourt Brace Jovanovich, 1976, pp. 183–183.
Gleason, Madeline. "Once and Upon." 1958. *The New American Poetry, 1945–1960*, edited by Donald Allen, Berkeley: University of California Press, 1999, pp. 125–127.
Grace, Nancy M. "Desert Journal ruth weiss (1977)." *Encyclopedia of Beat Literature*, edited by Kurt Hemmer, New York: Facts on File, 2007, p. 60.
Grace, Nancy M., and Ronna C. Johnson. *Breaking the Rule of Cool: Interviewing and Reading Women Beat Writers*. Jackson: University Press of Mississippi, 2004.
Khan, Yasmin. "The Desert Night." Poemhunter, 2013. Web: https://www.poemhunter.com/poem/the-desert-night/. Accessed 26 Jan. 2021.
Lee, Robert A. *The Routledge Handbook of International Beat Literature*. New York and London: Routledge, 2018.
Loy, Mina. "Mexican Desert." *The Dial*, vol. 70, 1921, p. 672.
Miles, Josephine. "Desert." 1934. *Poems: 1930–1960*. Bloomington: Indiana University Press, 1960, p. 15.
Monroe, Harriet. "A Garden in the Desert." *Poetry*, vol. 4, no. 6, 1914, p. 227. Web: www.jstor.org/stable/20570145. Accessed 26 Jan. 2021.
Moody, David A. *Tracing T. S. Eliot's Spirit: Essays on his Poetry and Thought*. Cambridge: Cambridge University Press, 1996.
MacKay, Marina. *The Cambridge Companion to the Literature of World War II*. Cambridge: Cambridge University Press, 2009.
Osborne, Sandra. "Iraqi Desert." Poemhunter, 2004. Web: https://www.poemhunter.com/poem/iraqi-desert/. Accessed 26 Jan. 2021.
Poe, Edgar Allan. "The Pit and the Pendulum." 1842. *Complete Stories and Poems of Edgar Allan Poe*, New York: Doubleday, 1966, pp. 196–207.
Shelley, Percy Bysshe. "Ozymandias." 1818. *The Norton Anthology of English Literature: The Major Authors*, Vol. 2, 10th Edition, edited by Stephen Greenblatt & M. H. Abrams, New York and London: Norton, 2018, p. 395.
Sherry, Vincent. *The Great War and the Language of Modernism*. Oxford: Oxford University Press, 2003.
Tucker, Brady Seth. "The Road to Baghdad." *Colorado Poets Center*, 2011. Web: https://coloradopoetscenter.org/poets/tucker_seth-brady/baghdad.html. Accessed 26 Jan. 2021.

weiss, ruth. *Desert Journal*. Boston: Good Gay Poets, 1977.
Wenders, Wim. *Paris, Texas*. Twentieth Century Film, 1984.
Wordsworth, William. "The Prelude." 1805. *William Wordsworth: Complete Works*, Hastings: Delphi Classics, 2013.

Chad Weidner
Reaching Towards the Light: Transitory Spaces and the Negated Material Body in Selected Texts by ruth weiss

ruth weiss is receiving increasing attention, and more study can bring new environmental understanding not just to core Beat texts but also to more outlying Beat-affiliated writers and the variety of their arts. An intriguing multicultural personality, ruth weiss resists easy or generic classification. She is not only a performance poet, since that devalues the allure of her textual forms. However, performance was essential to her literary and philosophical identity. weiss' position as Beat is now generally recognized, with caveats. She led a Beat lifestyle, wrote jazz-inspired poems, and embraced the Beat label, and it really does fit (Carden, 83). One aim of this paper is to highlight an internationally-oriented female Beat writer, ruth weiss, thus challenging some of the common views about the composition of the Beat Generation. A secondary goal is to bring Beat works into further dialogue with current environmental discourse. Indeed, ruth weiss' escape from Vienna and the Nazis was itself a response, of sorts, to the stresses and limits of material culture, which partly generated the ecological, economic, and social conditions for the Second World War. weiss contributed to a later international flourishing of Beat poetics. Recognition of this fact is important in acknowledging the greater transnational dimension to Beat Studies that has become so evident in recent years. However, important questions remain: To what extent can green criticism benefit by engaging unfamiliar and experimental transnational texts written by women? Can Beat studies be enhanced by environmental readings of unfamiliar texts by historically-neglected writers affiliated with the Beats? What follows, then, is consideration of representative excerpts from ruth weiss' *South Pacific* (1959), *Blue in Green* (1960), and *Desert Journal* (1977) to reveal ways the texts engage in environmental discourse. weiss' work explores transitory spaces as well as the negated material body. Consequently, this paper provides a green perspective on ruth weiss, who remains an eclectic polyglot, a transnational, cross-century Beat wanderer.

weiss has received modest attention from critics and scholars over the years, though given more recent interest in women Beats and the transnational aspects to the movement more generally, more attention is welcome.¹ Given her longevity and prolific poetic output, gaps in the scholarship on weiss remain curious. In *Can't Stop the Beat: The Life and Words of a Beat Poet* (2011), Horst Spandler emphasizes weiss' enigmatic use of language and unconventional literary structures as a possible reason her works lack complete critical engagement. He maintains that her "avant-garde language [...] eliminated conventional structures, and therefore was not easily accessible for a wider readership" and in part caused Beat publishing outlets to ignore her writing, though weiss acknowledged that her childhood compulsion to hide played an ongoing role (xviii). In a wider sense, Beat publishing outlets missed another opportunity, since weiss' work is relevant to contemporary environmental dialogue, is ideologically and directly politically motivated, and because weiss was such an incredibly productive Beat woman writer. However, Spandler makes a good point regarding the unconventionality of her work and her determined use of avant-garde language as a potential barrier to the wider reception of much of her work. Brenda Knight argued in 1996 that "ruth weiss is finally getting the attention she has long deserved"² (247), though this foreseen shift has not yet fully occurred. Matt Theado (2004) asserts that many of the women Beats, including Joanne Kyger and weiss "have been largely left behind in the last thirty years of Beat studies" (759). The discussion of weiss' exact position in the Beat canon is interesting and needs further study. Ronna Johnson argued that weiss and others occupied a shifting space within the Beat Generation movement and on the fringe of the literary avant-garde (10), though I resist this suggestion to some degree. What is clear is that while the publication of this book attests to growing interest in weiss, her position in the Beat canon remains unsettled.

weiss' German-language background was also a possible source of suspicion in the American reception of her work, as was her gender. The misogyny of the male Beats was echoed in the critical literature and popular press at midcentury. The Beat literary circle was largely white and male, and at least in the early formative stages of the movement was not particularly interested in collaborating with women. Strangely, one notable exception was traditional Beat adversary Herb Caen, who recognized weiss' potential, albeit very late. Caen coined the term Beatnik in 1958, which struck a chord in the popular consciousness and forever damaged the reception of the larger Beat project (Caen 1958). While Caen was a frequent opponent of the Beats in its formative period, he saw the

---

1 See also Carden (2018), Castelao-Gómez (2016), Grace and Skerl (2012), Thomson (2011), Grace (2004), and Knight (1996).
2 See also "'Beat Generation Goddess' S.F. Beatnik Scene Poet Ruth Weiss Set to Perform" (Eichler).

power of weiss' work, and in 1993 called her a "Beat Generation Goddess" (qtd. in Knight, 241). While there is more work to be done on the female Beats, with few exceptions, the Environmental Humanities have largely neglected the work of Beat writers. Grace and Skerl acknowledge the wider "environmental activism through poetics" (7) of the transnational Beats, though their 2012 collection does not really take the opportunity to explore the deeper ecological potential of the Beats beyond the suggestion. While it is recognized that Gary Snyder, Michael McClure, and even Ginsberg and Kerouac drew on the environment and nature as a basis for much of their writing, the Beat ecological circle has not yet been closed. This is particularly important now since the Environmental Humanities, which focus on the connections, tensions, and overlap between nature and culture, have made wide interventions into the study of culture. I worked on similar questions in *The Green Ghost: William Burroughs and the Ecological Mind* (2016), and green readings of the wider Beat canon seem a reasonable next step.

Academic fields are merging around climate crises, and the Environmental Humanities have developed deep and varied methods to examine culture and the environment. While the hard sciences have worked on ecological matters for a considerable amount of time, in the wider context the study and value of the environment in literature is undervalued. But humans have employed the creative cultural imagination to examine their relationships with nature for ages, and the study of culture has much to contribute. As Braidotti et al. (2013) assert:

> [T]he environmental humanities assume that modes of social belonging and participation are mediated by cultural representations and interpretations of them [...] The humanities can help us enhance changes in individual and social behaviour that promote sustainability. They do so by developing a better understanding of the cultural factors that construct the social imaginary, and so shape public representations of sustainability. This is achieved through the history and analysis of language, literature (ecocriticism), cultural images and representations in the arts and media, documentaries, films, computer games and Internet applications. (507)

This chapter suggests that ecological identity is shaped in part through contact with literature, with the contemporary understanding that climate change is real and humans contribute to the changing geology of the planet. Such changes will be evident in the geological record for millennia to come and have prompted many to employ the Anthropocene as a critical-concept.[3] What is clear and im-

---

[3] Donna Haraway prefers the more radically inclusive term "Chthulucene" (2016). Related neologisms include Capitalocene (Moore) and Pyrocene (Pyne). See also Nixon's *Slow Violence and the Environmentalism of the Poor* (2011).

portant here is to recognize that weiss' work has been neglected in some questers of Beat Studies, just as analysis of entire classes of literature and culture have been neglected by the Environmental Humanities.

Also interesting are the ways in which Beat Studies has opened itself up to acknowledging the contributions of writers to the construction of a broader and lasting planetary Beat ethos of environmentalism in poetic practice. Increasing transnational perspectives show that the Beat phenomenon was not merely an American product focused on the urban spaces of New York City, the Mountain west of Denver, and the left coast of San Francisco. One of the more interesting developments in Beat Studies has been the complex and transnational aspects of the movement, as the critical literature has shown this arc over the last decade. While progress has been made, there are still opportunities to develop new perspectives on the Beat Generation, and its greater contribution to postwar culture in North America, Europe, and elsewhere.[4] Thus, an environmental foray into the work of weiss, this chapter considers representative examples from three texts: "In a Japanese Tree-Garden" (1959), *Blue in Green* (1960), and *Desert Journal* (1977). These writings were selected not only because of their aesthetic qualities, but also because they have so far not benefited from enough research within in ecocritical circles. Looking at some representative examples will question to what degree weiss practiced a poetics of environmental praxis in interrogating transitory aesthetic spaces as well as exploring the radical philosophical potential of the negated material body.

"In a Japanese Tree-Garden" from *South Pacific* immediately provides readers with direct access to unmistakable environmental content. Published in San Francisco by Adler Press, the poetic content is explicitly ecological. Consider the following excerpt:

> flat is my ground and brown my soil
> i perch in flight
> i venture to the edge of things
> being rooted
> (n.p.)

This excerpt is interesting since it creates green images in the mind, advocates strength and resilience, and yet at the same time projects an active environmental imagination. It also provides a glimpse into the speaker's mind: "flat is my ground and brown my soil" can be read as implying a sort of anchoring to modest principles of the soil. The reader cannot know for sure if the speaker is really

---

4 See also Rod Phillips' *"Forest Beatniks" and "Urban Thoreaus"* (2000).

weiss, but the speaker feels rooted in the Anthropocene, an unfixed era of constant change, uncertainty, and even existential dread.[5] In "In a Japanese Tree-Garden," the speaker remains grounded, sturdied by the earth, by the simple basic material of the soil, the humblest material from which humans emerge and to which all people return. This is remarkable in thinking about the connections between writing and nature since the excerpt suggests that a solution to the postwar malaise in which the Beats found themselves immersed might not be a return to an escapist transcendent romanticism, but to more basic core elements. Yet, the speaker is not only rooted in the earth. In a strange incongruity, the speaker is both rooted and yet simultaneously exclaims: "i perch in flight," and "i venture to the edge of things / being rooted." This is particularly thought-provoking since it suggests aspiring towards something higher, more divine, perhaps even beatific, and that in itself is just so very Beat. Therefore, the piece delves into the unresolvable metaphysical questions that continually bedevil the human mind. What is the purpose of human existence? This enduring question speaks to the greater human need for self-actualization and understanding as well as the Beat endeavor to reach a higher state of consciousness.

In terms of an environmental reading, the content of "In a Japanese Tree-Garden" concerns contact with, and remaining cognizant of, nature. Later in the poem, weiss writes: "i spread and cluster / i yawn and smile / i know we are all here." There is something very benevolent happening here, and there are hints of kinship and acknowledgement of the sublime. James Hopkin says the Environmental Humanities are exciting since they do much more than summon the vigor of some sort of transcendental distraction, and instead engenders recognition of the larger circle of life to which we all belong. weiss does this throughout this poem and in her work more widely, hinting at the "inter-relatedness of all factors within the ecosystem [...] to the phenomena of the natural world" (11). In this way, weiss works almost as a proto-ecocritic, utilizing inventively poetic allusions and regenerative language that speak to the core of environmental thought, which include the recognition of literature as a source of redemption and potential cultural recovery. In a further extract from "In a Japanese Tree-Garden," weiss again utilizes a rhetorical strategy of paradox to highlight issues of ecology:

> awareness
> where lights
> and darks

---

5 Early appraisals of the Beats understandably did not employ the Anthropocene as a critical concept, though they did speak of similar existential concerns. See also Mailer (1957).

> dart
> teasing
> we stem
> in togetherness

At first the reader might be slightly puzzled by the contrast of lights and darks here, but what is so interesting about this, and the wider Beat project, is the recognition of the totality of the human experience set in the word 'beatific'. Life is both revelatory and distressing, and so is nature. "[W]e stem / in togetherness" also advocates the ongoing development of shared experience and civic responsibility, which are necessary prerequisites to biocentric thinking, the idea that everything is connected, and by extension, that humans have real agency and can take ownership and responsibility by developing new forms of civic engagement. In other words, this extract can be read as subtly advocating the development of shared accountability and stewardship, which can rightfully be called a precondition for ecological citizenship.

Blue in Green also captivates the green imaginary. *Blue in Green* is an unusual piece, even in the context of weiss' wider work, and it resists the idea of genre. *Blue in Green* is not really prose, not really, but it is also not a poem, at least not what one would normally expect a poem to be or look like. Instead, *Blue in Green* resembles a long poem printed over 12 thick pages. Consider the following excerpt:

> green thoughts
> in birth
> in cut-hurt wild-willed
> greenfulness
> the whisper-chant
>
> green thoughts
> a-whirr
> (1)

This selection resists generic reading, even a basic green reading, though further deliberation allows space for the environmental imagination. The shape of *Blue in Green*, as suggested, is unusual. The repetition of "green" here conjures up loose metonymic notions of nature. The repeated usage of "green" can also be considered a form of repetend or refrain, which is the recurring or irregular use of a word for poetic effect. The function of the repetend here infers generic thoughts of nature, though the reader cannot be totally sure. The lines "in birth / in cut-hurt" (1) can be understood as signifying the trauma of mammalian birth. The other lines remain obscure, though the image of "the whisper-chant" appeals to the auditory, and the "green thoughts / a-whirr" (1) speak to the human mind. Matt Theado argues that "weiss's spontaneous method of free association demands a sympathetic reader, one who is willing to free associate and to swing with both meaning and message" (759). This is key to reading weiss, since she relies on the participation of the reader to fill in the blanks. What can also be seen here and elsewhere in the work of weiss is recognition of the earth, of the primeval, of the human senses, of birth. A green reading here cannot provide a conclusive message, but the writing forces readers to think, which might best be described as a key function. While making readers think cannot directly be tied to environmental discourse, it does serve a didactic purpose, as does applied theatre. If good writing makes people rethink what literature is and does, then they might by extension reconsider unsustainable practices today. The final text for consideration in this chapter encourages embrace of the void, and by extension the current human predicament, to capitalize on the ephemeral human moment.

Desert Journal was published in 1977 by Good Gay Poets in Boston. Artist Paul Blake intertwines this collection of 40 poems with a number of drawings that are evocative of prehistorical petroglyphs, of the thunderbird and other prehistoric dreamlike-images that dot the rugged landscape of the west of America, harkening back to unknown and unfamiliar cultures which developed profound understanding of the value of nature, and of cultural representations of nature. In the context of weiss' wider work, *Desert Journal* has benefited from critical attention. This could be because *Desert Journal* functions as an artistic and philosophical declaration or manifesto. *Desert Journal* is laden with exciting poems. Much existing criticism on the book emphasizes the visual aspects to weiss' poetry here. Nancy Grace says that weiss understood that she "is often called a visual poet [which is not] surprising since she has long been immersed in the visual arts" (58). However, what can also be seen in *Desert Journal* is much more than just suggestions of a more primitivist and creative humanist philosophy and an interest in visual art. More than that, the collection harkens back to the historical avant-garde, including "Tzarian dadaism" as well as "Beat jazz and performance practices drawn from Romantic and Buddhist belief in 'first thought is best thought'"[6] (Grace, 59). Grace also sees weiss' collection as "frequently collaborative, conjoining words with music in public performance to create a dialogue [...] with musicians and the audience" (59). The end result exists somewhere "on the cusp of modernism and postmodernism" (60). In this way, too, one can suggest that weiss fulfills another measure of sorts for Beat aesthetic affiliation, since

---

[6] "First thought, best thought" as a creative-rhetorical concept is open to debate.

Beat works often exist on the spectrum between Modernism and Postmodernism.[7]

Desert Journal is broken down into forty days of poems, which resembles a sort of vision quest, or solitary mission of scope and great importance to forming one's own identity. The suggestion of a vision quest implies reflection and spiritual communion with nature in both the tradition of Native American cultures and the ancient Western ritual of solitary journey into the desert based on faith. What can also be found in the collection are a number of unexpected digressions which are playful, unconventional, and dadaesque. Consider the following excerpt:

> Third Day:
>
> SOUR SAUCE
> SOUR SAUCE
> MARINATE THIS LIFE
> WITHOUT SEEMING CAUSE
>
> VICE HAS LOST ITS SHARP
> SHARP HAS LOST ITS SPICE
> VICE HAS LOST ITS SHARP
> HARP HAS LOST ITS STRING
> STRING HAS LOST ITS WING
> HARP HAS LOST ITS SPRING
> SPRING HAS LOST ITS SING
> WING HAS LOST ITS BIRD
> BIRD HAS LOST ITS TURD
> VICE HAS LOST ITS SPICE
>
> AND TURD ON THE THIRD
> DAY HOPE AS SPICE THE NIGHT
> NIGHT HAS LOST ITS FIGHT
> TO BE THE ONE UNKOWN
> WANT TO BE
> AS SOWN THE SEED
> THE NEED – – –
> (n.p.)

What can be found here, and throughout *Desert Journal*, is an intriguing and compelling employment of repetition, rhyme, off-rhyme, and especially language play as rhetorical practice. Robert Bennett (2005) emphasizes the strategies employed by weiss as corresponding with performative practices of other Beats.

---

[7] "Postmodernism" as a critical term in the Beat context is also contentious. See also Harris (2006) and Loranger (1999).

Bennett observes, "Ruth Weiss developed a sense of linguistic play that is as performative [...] as Kerouac's and Ginsberg's" (183). This is a rather remarkable claim and has value given the centrality of performance to the larger Beat poetic project. Linguistic play is evident in "Third Day" through the repetition of rhyme, sounds, and of the earthy and even scatological vocabulary, including the "BIRD HAS LOST ITS TURD" and "TURD ON THE THIRD". Playful engagement with the vulgar also parallels the wider Dada rejection and active subversion of mainstream discourse.[8] Language play is further exemplified by weiss' refusal to capitalize her name, which violates German grammatical and linguistic codes. Another example from *Desert Journal* speaks to both transitory spaces and the negated material body. Consider the following extract from "Thirty-Second Day," which is displayed over several pages:

> A
> VOID
> CANNOT
> BE
> AVOIDED
> (n.p.)

The capital letters here at first suggest solemnity or invoke a message of deep caution, but further reflection suggests more intriguing links to environmental thinking. When thinking about the purpose of such a powerfully ideological reading of "A VOID / CANNOT / BE / AVOIDED," one cannot help but consider Buddhist teaching about the recognition and embrace of the emptiness, and of the deep Ginsberg-inspired void (see Bellarsi). When contemplating the ecological significance of the idea of the void in the context of literature, William Rueckert, the first known person to use the term "ecocriticism" in his essay "Literature and Ecology" (1978b), warns: "History is not a nightmare [...] but a corpse, a waste land, a void [...] a world in which there are no sanctuaries, a sterile and impotent self, an ontological cemetery, a catalogue of dead visions and dreams. Revolution, Transcendence, Immersion, Cognition, and ironic resignation" (1978a, 63). Rueckert seems to accept the evolutionary dead-end ahead. Darwin has shown that there are so many more paths to death than to life, and the fossil record verifies this fact. However, Rueckert's caution is not just defeatist. Recognition of the power of the void can work directly towards environmental praxis in providing immediacy and urgency to culture, as can be seen in the work of weiss.

---

8 See also Zobel.

An accepted theory called "heat death of the universe" postulates that once the capacity of the universe to generate new galaxies fails, eventually a point of total heat death will occur, which means not only absolute extinction of all creatures everywhere, but the end of all movement everywhere, a forever frozen void.[9] While a deeply disturbing prospect that both frightens and troubles humans, this knowledge that people exist in a temporary and transitory space also provides opportunities to reconsider human time on the earth. And this is where weiss' poetry and Beat writing really works towards conservational praxis. By acknowledging the limited time, people have the chance to imbue their lives with meaning. This is what Heidegger (1927) also believed really separates humans from other animals. If humans are doomed, and if they know it, then perhaps it is simply best to accept that the void cannot be avoided. Interesting also is that acceptance of the void appears throughout the work of the wider Beat Generation, including in Corso's "Bomb" (1958) and Burroughs' *The Western Lands* (1987).

What also can be found in weiss' *Desert Journal* is not simply an acknowledgement of the impossibility of ever escaping the void, but existence in a transitory space, and the negated material body. Nancy Grace argues that *Desert Journal* unmistakably demonstrates "that for weiss language is the vehicle for individual escape from [...] materiality. The poem suggests that to escape the materiality of the human body and the desert, the poet must acknowledge language as both the enemy of escape and the tool of escape" (63). This is essential to understanding the wider Beat project. Preston Whaley asserts in *Blows like a Horn* that Beat writers' adhering to the aesthetic potential of Kerouac's "Spontaneous Prose" ultimately allowed them to break down distinctions "between [...] bad thought and good thought, to the larger parameters of sinner and saint, mind and body, enemy and friend, civilization and ecology" (35). Furthermore, current materialist discussions challenge "the classical ontologies of mind/body and self/world dualism" (Connolly, 399). The advantage of material studies is the willingness to question long-lasting dualisms such as culture vs nature, technology vs nature, mind vs body, and the assumed materiality of the human.[10] Such a shift in thinking is important in a time when ecological calamities force humans to intellectual and creative limits. Acknowledgement of the negated material body also allows space for thinking beyond the self. Additionally, the concept of movement as artistic expression remains central to weiss. Mary Carden says that in nomadism, weiss "constitutes identity in/as movement" (85). What can

---

9 See also Adams 1997.
10 See also Marx (1964) and Braidotti (2013).

be found in *Desert Journal* is a link between autonomy and authenticity with the endless journey, which recalls one of the central principles of the wider Beat project.

Furthermore, weiss interrupts mainstream language and expectations of what poetry is and is supposed to do. weiss' wider work is evocative of Burroughs' writing, as well as Dada and Surrealist fracturing of conventional reader expectations. It manifests itself in weiss' writing in the rhetorical strategies of language play, which while evident in her work, also resists a straightforward first-wave ecocritical reading. It seems that often no clear environmental message seems plausible here. weiss shares the same trick here as William Burroughs: the active countering of reader expectations and associated hegemonic thinking about the form and function of literature. Countering mainstream discourse is a common Beat strategy on the more experimental fringes of the movement. Language play in the work of weiss also echoes the urgent Dada call to action to save society from unimaginable industrial-scale death. The appropriation of discourse is highly provocative, but also speaks to the potentials of dismantling and rebuilding language.[11] This composting of language can be seen elsewhere throughout the Beat canon, particularly of linguistic montage of texts in *Minutes to Go* and Burroughs' *Cities of the Red Night*, *The Place of Dead Roads*, and *The Western Lands*. Irene Gammel and John Wrighton assert that the remnants of language are valuable cultural material, as "language itself is a cultural litter to be recycled and renewed [and is] subject to the ecological laws of decomposition and recomposition" (10). In the context of weiss' wider corpus, one could say that the combination of urgent subversion of mainstream discourse and the environmental content of much of her work together pair for an interesting combination that highlight the emptiness of consumer culture in the second half of the twentieth century, but at same time speaks to the larger Beat pursuit of ecstasy. Read in this way, weiss' art can be seen as occupying a transitory space between performance art, the Beat Generation, and the historical Avant-Garde. At the same time, her work speaks to more ghostly questions including the negated material body, or at least the potential of humans to escape their physical limitations, which revises the longstanding mind vs body debate. By investigating transitory spaces of language, cultural identity, literature, and poetry, weiss participates in a kind of disembodied "phototropism," which is the tendency of plants and other microscopic organisms to favorably respond to sunlight, or to move towards or grow towards the sunlight. And in this way, it

---

11 See also Tarlo (2009).

can be said that weiss speaks directly to larger the Beat pursuit of blessed enlightenment, civic responsibility, and community.

Works Cited

Adams, Fred C. and Gregory Laughlin. "A Dying Universe: The Long-Term Fate and Evolution of Astrophysical Objects." *Reviews of Modern Physics*, vol. 69, no. 2, 1997, pp. 337–372. Web: doi:10.1103/RevModPhys.69.337. Accessed 15 Apr. 2018.

Beiles, Sinclair, et al. *Minutes to Go*. San Francisco: City Lights Books, 1960.

Bellarsi, Franca. "Confessions of a Western Buddhist 'Mirror-Mind': Allen Ginsberg as a Poet of the Buddhist 'Void.'" PhD thesis. Université Libre de Bruxelles, Faculté de Philosophie et Lettres, Bruxelles, 2002.

Bennett, Robert. "Deconstructing and Reconstructing the Beats: New Directions in Beat Studies." *College Literature*, vol. 32, no. 2, 2005, pp. 177–84. Web: http://www.jstor.org/stable/25115273. Accessed 15 Apr. 2018.

Braidotti, Rosi. *The Posthuman*. Cambridge: Polity Press, 2013.

Braidotti, Rosi et al. "The Humanities and Changing Global Environments." *World Social Science Report 2013: Changing Global Environments*, Paris: OECD Publishing/Unesco Publishing, 2013, pp. 506–8.

Burroughs, William S. *Cities of the Red Night*. New York: Picador, 1981.

Burroughs, William S. *The Place of Dead Roads*. New York: Henry & Holt, 1983.

Burroughs, William S. *The Western Lands*. New York: Viking Penguin, 1987.

Caen, Herb. "Pocketful of Notes." *San Francisco Chronicle*, April 2 1958. Web: https://www.sfgate.com/news/article/Pocketful-of-Notes-2855259.php. Accessed 15 Apr. 2018.

Carden, Mary Paniccia. *Women Writers of the Beat Era: Autobiography and Intertextuality*. Charlottesville: U of Virginia Press, 2018.

Castelao-Gómez, Isabel. "Beat Women Poets and Writers: Countercultural Urban Geographies and Feminist Avant-Garde Poetics." *Journal of English Studies*, vol. 14, 2016, pp. 47–72. Web: http://doi.org/10.18172/jes.2816. Accessed 15 Apr. 2018.

Connolly, William. "The 'New Materialism' and the Fragility of Things." *Millennium*, vol. 41, no. 3, 2013, pp. 399–412. Web: https://doi.org/10.1177%2F0305829813486849. Accessed 15 Apr. 2018.

Corso, Gregory. *Gasoline*. San Francisco: City Lights, 1958.

Eichler Network. "'Beat Generation Goddess' S.F. Beatnik Scene Poet Ruth Weiss Set to Perform in Jazz Setting at Berkeley Gallery." *Eichler Network*, 2016 Web: https://www.eichlernetwork.com/article/beat-generation-goddess. Accessed 15 Apr. 2018.

Fischer, John. "The Editor's Easy Chair: The Old Original Beatnik." *Harper's Magazine*, April, 1959, pp. 14–16. Web: https://harpers.org/archive/1959/04/the-old-original-beatnik/. Accessed 1 Oct. 2018.

Gammel, Irene, and John Wrighton. "'Arabesque Grotesque': Toward a Theory of Dada Ecopoetics." *Interdisciplinary Studies in Literature and Environment*, vol. 20, no. 4, 2013, pp. 1–22. Web: www.jstor.org/stable/44087293. Accessed 15 Apr. 2018.

Gold, Herbert. "How to Tell the Beatniks from the Hipsters." *Noble Savage*, Spring, 1960, pp. 132–39.

Grace, Nancy. "Ruth Weiss's *Desert Journal:* A Modern-Beat-Pomo Performance." *Reconstructing the Beats*, edited by Jennie Skerl, New York: Palgrave Macmillan, 2004, pp. 57–71.

Grace, Nancy M., and Jennie Skerl. *The Transnational Beat Generation*. New York: Palgrave Macmillan, 2012.

Haraway, Donna Jeanne. *Staying with the Trouble: Making Kin in the Chthulucene*. Durham: Duke University Press, 2016.

Harris, Oliver. "Not Burroughs' Final Fix: Materializing the Yage Letters." *Postmodern Culture: An Electronic Journal of Interdisciplinary Criticism*, vol. 16, no. 2, 2006. Web: http://doi.org/10.1353/pmc.2006.0017. Accessed 30 Jan. 2021.

Heidegger, Martin. *Sein und Zeit*. Halle: Niemeyer, 1927.

Hopkin, James. "In the Green Team: James Hopkin Looks at How Eco-Critics Are Sending Ripples through Literature." *Guardian*, May 12, 2001. Web: https://www.theguardian.com/books/2001/may/12/scienceandnature.highereducation. Accessed 18 Apr. 2018.

Johnson, Ronna. "Anatomizing the Beat Generation." *Breaking the Rule of Cool: Interviewing and Reading Women Beat Writers*, edited by Nancy M. Grace and Ronna C. Johnson, Jackson: UP of Mississippi, 2004, pp. 3–42.

Knight, Brenda. *Women of the Beat Generation*. Berkeley: Conari, 1996.

Loranger, Carol. "'This Book Spill Off the Page in All Directions': What Is the Text of *Naked Lunch?*." *Postmodern Culture*, vol. 10, no. 1, 1999. Web: http://doi.org/10.1353/pmc.1999.0033. Accessed 18 Apr. 2018.

Marx, Leo. *The Machine in the Garden: Technology and the Pastoral Ideal in America*. New York: Oxford University Press, 1964.

Mailer, Norman. "The White Negro (Superficial Reflections on the Hipster)." *Dissent*, August 14, 1957. Web: https://www.dissentmagazine.org/online_articles/the-white-negro-fall-1957. Accessed 15 Apr. 2018.

Moore, Jason W. "The Capitalocene, Part I: On the Nature and Origins of Our Ecological Crisis." *The Journal of Peasant Studies*, vol. 44, no. 3, 2017, pp. 594–630. Web: https://doi.org/10.1080/03066150.2016.1235036. Accessed 15 Apr. 2018.

N.N. "Beatniks Just Sick, Sick, Sick." *Science Digest* July, vol. 46, 1959, pp. 25–26.

N.N. "Bye, Bye, Beatnik." *Newsweek* July, vol. 1, 1963, p. 65.

N.N. "Every Man a Beatnik?" *Newsweek*, vol. 53, 1959, p. 83.

Nixon, Rob. *Slow Violence and the Environmentalism of the Poor*. Harvard UP, 2011.

Phillips, Rod. *"Forest Beatniks" and "Urban Thoreaus": Gary Snyder, Jack Kerouac, Lew Welch, and Michael McClure*. New York: Peter Lang, 2000.

Pyne, Stephen. "Big Fire; or, Introducing the Pyrocene." *Fire*, vol. 1, no. 1, 2018, p. 1. Web: https://www.mdpi.com/2571-6255/1/1/1. Accessed 18 Apr. 2018.

Rueckert, William. "Into and out of the Void: Two Essays." *The Iowa Review*, vol. 9, no. 1, 1978a, pp. 62–71.

Rueckert, William. "Literature and Ecology: An Experiment in Ecocriticism." *The Iowa Review*, vol. 9, no. 1, 1978b, pp. 71–86.

Spandler, Horst. "ruth weiss and the American Beat Movement of the '50s and '60s." [Introduction] *Can't Stop the Beat: The Life and Words of a Beat Poet*. By ruth weiss. Studio City, CA: Divine Arts, 2011, pp. IX–XXVI.

Tarlo, Harriet. "Recycles: The Eco-Ethical Poetics of Found Text in Contemporary Poetry." *Journal of Ecocriticism*, vol. 1, no. 2, 2009, pp. 114–13.

Theado, Matt. "Beat Generation Literary Criticism." *Contemporary Literature*, vol. 45, no. 4, 2004, pp. 747–761. Web: www.jstor.org/stable/3593550. Accessed 15 Apr. 2018.

Thomson, Gillian. "Gender Performance in the Literature of the Female Beats." *CLCWeb: Comparative Literature and Culture*, vol. 13, no. 1, 2011, pp. 2–8. Web: https://doi.org/10.7771/1481–4374.1710. Accessed 15 Apr. 2018.

Weidner, Chad. *The Green Ghost. William Burroughs and the Ecological Mind*. Carbondale: Southern Illinois University Press, 2016.

Whaley, Preston. *Blows Like a Horn: Beat Writing, Jazz, Style, and Markets in the Transformation of U.S. Culture*. Cambridge, MA: Harvard UP, 2004.

weiss, ruth. *South Pacific*. San Francisco: Adler Press, 1959.

weiss, ruth. *Blue in Green: [Poems]*. San Francisco: Adler Press, 1960.

weiss, ruth. *Desert Journal*. Boston: Good Gay Poets, 1977.

weiss, ruth. *Can't Stop the Beat: The Life and Words of a Beat Poet*. Studio City, CA: Divine Arts, 2011.

Zobel, Greg. "Ruth Weiss: Making a Deeper Groove in Europe." *The Kerouac Connection*, vol. 29, 1998, pp. 31–33.

Hannes Höfer
Traditionally New: The Jazz & Poetry Work of ruth weiss

Introduction

ruth weiss was one of the first artists, if not *the* first artist, who performed her poetry together with jazz musicians in San Francisco in the 1950s. She thus established a genre that became popular as a key element of the Beat Poetry Movement (Antonic 2018b). That being said, however, she failed to ever receive the degree of recognition she truly deserved. The first Jazz & Poetry recordings were produced by writers like Kenneth Rexroth, Lawrence Ferlinghetti, or Jack Kerouac who took credit for inventing Jazz & Poetry for themselves without mentioning ruth weiss at all (Whaley, 64–70). They also failed to credit forerunners like Langston Hughes or Pierre Reverdy, who recorded the first known performance of poetry together with jazz music in France in 1937 (Schroeder and Miller, 645). The Jazz & Poetry experiments of the Beat writers had mostly ended by the early 1960s. ruth weiss, however, continued her collaboration with jazz musicians. Like the Beat poets, weiss preferred the creativity, spontaneity, and liveliness of bebop, and shared their emphasis on the sound and oral dimensions of poetry. Less obvious than in Beat aesthetics, the search for transcendence was not as central in weiss' poetry as was the case in poetry of authors such as Kerouac (Weinreich, 51–54). And weiss was never really interested in promoting herself as a solitary genius (Carden, 91). Instead, she always recognized the importance of creative collaboration. It is precisely this type of mutual artistic enrichment that was embodied in weiss' Jazz & Poetry work.

This chapter analyzes and compares three Jazz & Poetry recordings of ruth weiss with various other jazz musicians in order to study the specific way in which the poet collaborated with them. I will start with a home recording of the poet performing from her volume *Desert Journal* together with bass player Benfaral Matthews in 1969. I will discuss the performance of the poem "First Day" from this recording because it provides a comprehensive insight into the collaboration between weiss and Matthews and it allows for a comparison with a recording of the same poem with the bass player Daney Dawson fifteen years later. This comparison will demonstrate that weiss was able to act and react versatilely in collaborations with different musicians in order to create different versions of the same poem. The last example studied in this chapter is from a 2018 live performance with musicians Rent Romus, Doug O'Connor, and Hal Davis. This performance reveals how ruth weiss and the musicians combined her own biography with respect to the history of jazz in order to create a performance that was simultaneously contemporary in nature while also demonstrating an awareness of the historic development of Jazz & Poetry. Before taking a closer look at these performances, the next section foreshadows some of the general issues that need to be taken into account when analyzing the specific combination of words and music in a Jazz & Poetry performance.

Jazz and Poetry

In general, "Jazz & Poetry" or "poetry and jazz" denotes the performance of poetry in combination with jazz music. It can be considered a sub-genre of "jazz poetry" (Feinstein, 61–88), as "a kind of synesthetic minor aspect" of jazz in fictional and non-fictional literature (Hunkemöller, 1431), in addition to being part of the history of jazz in general (Schroeder and Miller, 643–52), and of both the Black American musical culture (Jones) and the Beat movement in particular (Hrebeniak). In any case, for the scope of this chapter, one can benefit from an understanding of the different angles by which the combination of text and music is often studied. Research tends to favor intermedial or performance studies methods, depending on the existing material: Do we have audio and/or video recordings, or are we talking about a live performance experience? The live experience or the video recordings allow for a comprehensive and transparent analysis of Jazz & Poetry performances because we can then consider the positions of the artists on stage, their interaction or gestures, and other performance elements that shape the combination of text and music (Jones, 19). Unfortunately, in most cases we lack video footage and therefore have to rely on the audible aspects of performance. In consequence, most research on Jazz & Poetry focuses on the intermedial dimension of the combination of text and music. In this respect, scholars favor the intermedial adaptation as an ideal combination, which means a mutual influence of text and music and an equal juxtaposition of both genres in order to produce a surplus of significance that is higher than the single significance of both genres involved (Jackson). Frieder von Ammon has presented the latest and most comprehensive analysis of Jazz & Poetry projects in his study on the jazz poetry of the Austrian poet Ernst Jandl, which is highly relevant for the approach employed in this chapter.

Based on Werner Wolf's theory of musico-literary intermediality (Wolf), von Ammon considers Jazz & Poetry performances as most convincing when they achieve a mutual interaction of poet and musician(s) that results in the adaptation of musical structure in text and vice versa (von Ammon, 354). This appreciation of spontaneity and interaction on equal terms, which is fundamental to the theory of intermediality, is similarly important to jazz aesthetics, where equal communication in a live performance situation is one of the most important values. The emphasis on interaction leads to an equality of the two arts involved in Jazz & Poetry. However, emphasizing interaction simultaneously leads to a hierarchy of the different forms of Jazz & Poetry. It rejects Jazz & Poetry projects such as the one by Joachim-Ernst Behrend in which he combined the recitation of poems of Heinrich Heine or Gottfried Benn with recorded jazz of Attila Zoller or Jay Jay Johnson. If we want to avoid this hierarchy, we might rather shift our focus from the adaptation of textual and musical structures to the mutual reinterpretation of text and music. In this respect, Frieder von Ammon can provide a guide to a certain extent.

After identifying the intermedial interaction of text and music as pivotal dimension of Jazz & Poetry, von Ammon analyzes the general impression of Jazz & Poetry performances. This impression can vary even in different recordings of the same text, as von Ammon convincingly demonstrates regarding three versions of Jandl's poem "etüde in f" ("étude in f"), which is performed on different occasions "as choral étude for (mezzo-)soprano, as parody of a sermon, and as veristic sound painting" (von Ammon, 404; author's translation). While the different setting alters the identical text into three different versions, this does not necessarily mean that one version is superior to the other (and von Ammon notes that all three versions share the same level of felicitous adaptation). It does reveal, however, the variety of meaning and the surplus of significance that can be produced when poetry is performed together with jazz. For this reason, it is worth focusing more intensely on aspects of textual and musical meaning. It thus behooves us to employ the same thoroughness in analyzing the level of textual and musical meaning as we do on the level of intermedial relation of structural analogies.

Analyzing the mutual reinterpretation of text and music leads to the question of what elements of text and music should be compared. Jackson's study on the collaboration of Amiri Baraka, David Murray, and Steve McCall on the 1981 record *New Music – New Poetry* provides some insights (Jackson, 364–7). He focuses on the development of tradition and therefore highlights the textual and musical references in relation to literary and musical traditional contents and forms. This provides an apt method for studying the interaction of meaning in Jazz & Poetry give that, just as words have meanings and connotations, so do literary forms. For example, generally speaking, it is commonly understood that a sonnet "mean" something different than free verse, and we can use this knowledge to describe "meanings" by reconstructing the cultural contexts and history of different forms of poetry. The same logic can be applied to music. Although single notes or a sonata form might not clearly "mean" something, specific musical styles like ragtime or free jazz always have a meaning that depends on a specific (socio-)cultural and historical context. To put it simply, while it is certainly complicated to identify the one meaning of bebop, one cannot deny that bebop is "different" than polka, and that connotations like "freedom" or "spontaneity" are closer to our understanding of the word "bebop" than to others like "luxury" or "pizza". Of course, and generalities aside, we still have to decide on a case-by-case basis if contradictory connotations like "liveliness" or "artificiality" are linked to bebop in a given example. This brings us to the Jazz & Poetry collaborations of ruth weiss.

The central question I want to address is how jazz music influences ruth weiss' recitation or performance of a poem. Does it highlight hidden or covert meanings of the poem? Or, alternatively, does it establish a new level of meaning that was not part of the poem beforehand? The emphasis on the text here does not mean that the analysis of text is more important than the analysis of music. As mentioned above, both points of view have to be considered equally in order to describe their interaction. However, weiss is, first and foremost, a poet. In this vein, my analysis will begin with the written text – with the study of her written poems – to later compare this with weiss' performances together with musicians.

Jazz & Poetry Recordings of ruth weiss

ruth weiss and Benfaral Matthews (1969)

As an unfortunate matter of fact, the Jazz & Poetry performances of ruth weiss in San Francisco in the 1950s are not recorded. However, we can get a pretty clear idea of her energy and high level of interaction if we listen to the 1969 recorded collaboration between ruth weiss and the bass player Benfaral Matthews.[1] Matthews, an assertive player with a sonorous sound and an accentuated rhythmical style, invites weiss to a strong (re-)action as she performs with big variation in pace, volume, and amplitude.

A good example of this dynamic is their version of "First Day" of the cycle of poems *Desert Journal* which deals with the arrival in the desert. The poem presents the desert as a threat but also as a space of new possibilities and experiences. Most threatening in this space is the fact that the threat cannot be defined. It

---

[1] Unreleased recording from the private archive of ruth weiss on reel-to-reel, digitized by and courtesy of Thomas Antonic.

appears in a line of tables that "were dropping into the cavity of mouth / slowly" (weiss 1977, "First Day")[2] and a bird that turns into the "fable-bird" of Poe's "The Raven". This bird is not a nameless threat in the poem, but is itself terrified, not knowing in which direction to go and "alone now / but still with feet / could land" (1977, "First Day"). Although the bird does not really know which way to go, the speaker wants to follow, trusting poetry and metrical feet would allow her to stay in touch with solid ground. The speaker learns from the bird that "all animal lives against" but that this is a "possibility of become" (1977, "First Day"). She decides not to be afraid of the unknown but to be empathetic with it or, as is repeated three times, ends up "loving the enemy more" (1977, "First Day"). Eventually, the speaker identifies with the bird, accepts her situation in the desert, and discards her friends' advice: "and the first day / is the worst day / in the desert // they tell one to make a fast last / slow—like a dirty bird" (1977, "First Day").

In a few words, the poem presents self-assertion obtained through empathy with the threatening unknown. The combination of threat and empathy guides the performance of weiss and Matthews, as each contribution expresses the threatening dimension of the poem while their sensitive interaction demonstrates the dimension of empathy. weiss varies her delivery. She starts slowly and firmly, setting small pauses between single words but quickly turns to a more breathless impulse resting on some vowels without losing her rhythmical strength. In the last two lines, she reverts to a slower pace, keeping the broad range she was using beforehand. Hence, the poet creates the impression of great urgency, which is not only caused by her delivery alone, but also by her reaction to Benfaral Matthews' playing.

Matthews starts with long deep tones on the double bass by using a bow. He creates a sense of menace, which intensifies when weiss starts speaking in a low voice. Although Matthews only plays sounds that are long in duration, he nevertheless pushes weiss to a faster and more intense delivery. This leads Matthews to establish a beat around 120 bpm by playing quarter notes, thereby acting as a pace-maker for weiss. Both now start playing around with the length of pauses and phrases but don't break the beat. Matthews goes on with some flageolet phrases, finally finding a walking bass line that he turns to a provisional end as weiss repeats the line "the seconds all felt the duel". He hesitantly enters again and is forced to a more rhythmical playing by weiss who skips the line "lightning struck" to combine the two nearly identical lines "do you REALLY

---

2 As the first edition of *Desert Journal* has no pagination all following quotes–if not noted otherwise–are from "First Day".

WANT YOUR COMPENSATION" / "DO YOU REALLY WANT YOUR COMPENSATION", which she speaks with a strong crescendo. Matthews starts again a walking bass and weiss adapts to the 4/4-beat in the lines consisting only of monosyllabic words. Matthews alertly reacts and skips the downbeats, thus creating a perfectly swinging clockwork together with weiss. As weiss again starts to play with stresses and pauses, Matthews surrounds the text with more melodic movement in eighth notes. In the end, weiss and Matthews finish laconically. weiss signals the end with a long and high "slow", which is followed by a calmly spoken "like a dirty bird" accompanied by a fading descending bass line. In sum, Matthews uses transparent phrases that seem to be chosen from a book of classical bass etudes. His deep tones and his sometimes-raspy bowing emphasize the feeling of threat that the poem expresses. However, this impression is not merely caused by his playing, but rather by the interaction with weiss. She and Matthews create a performance of the poem "First Day" that sounds like a threatening piece of chamber music played with enthusiasm. In this performance, they realize the meaning of the text (self-assertion obtained by empathy vis-à-vis a nameless threat) by collaborating in a very specific way. Looking at the bass accompaniment and the delivery separately, both emphasize the dimension of threat. It's the highly spontaneous and playful interaction of weiss and Matthews that stresses the dimension of self-asserted empathy. As such, both poet and musician do not transpose the textual meaning in elements of recitation and music alone, but in the performative interaction. Their ensemble play expresses an empathy necessary to cope with unknown menaces suggested in the poem.

ruth weiss and Daney Dawson (1984)

ruth weiss' version of "First Day" together with Daney Dawson is also a home recording, and although it is seven seconds shorter in length than the one with Benfaral Matthews, it gives the impression of being slower and calmer. weiss' voice has become a little bit deeper over the years and she speaks with less range and less rhythmical pauses within the phrases. Thus, her delivery sounds like a calm reading. She sticks to the original text with only one exception (see below). Her calm and confident voice supports the main theme of the poem – that one can obtain self-assertion through empathy vis-à-vis a nameless threat. Whereas weiss' delivery in the recording with Matthews is more urgent and agitated, here it is steadfast. A look at the musicalization reveals even more differences between the two performances.

Dawson also plays double bass but, unlike Matthews, without a bow. She accompanies weiss with a restrained walking bass spiced with some slides and falls. Here, the word "accompany" is true in the full sense of the word. In this performance, Dawson's playing follows weiss' voice. When weiss ends a line and pauses to give some space for Dawson, she terminates her musical phrase, too, and waits for weiss to start again. Only one exception to this pattern occurs; interestingly enough, this happens at the end of the poem, when weiss recites "slow—like a dirty bird", emphasizing the last syllable, which leads Dawson to play an ascending scale. weiss repeats the last line fast and dry and Dawson slides down the scale again to finish the performance. Although this passage is an exception from the written text, the musicalization here is typical for the whole performance. Unlike Matthews did with his strong beat, Dawson does not challenge weiss with her own ideas to change her delivery. Nevertheless, the fact that Dawson mainly reacts to the poet's pace and intonation does not mean that she accompanies weiss in an uninspired way. When weiss recites "as thunder BEFORE lightning struck / did anyone count?", Dawson imitates the "strike" with a loud and high sliding tone and echoes this sound a little bit deeper and quieter after the word "count". She repeats this effect when weiss repeats "lightning struck" some lines later. Thus, Dawson listens very sensitively and is highly aware of both the meaning of the words and the delivery of the speaker. She continuously finds simple but convincing ways to incorporate the text into her music. However, the performance doesn't sound as compact as the one with Benfaral Matthews. One reason might be that weiss and Dawson do not establish a continuous beat, as Dawson always waits for weiss to set the pace. While weiss and Matthews are pushed forward by the beat, weiss and Dawson are interacting very closely without any steady beat shifting from moments of tension to pauses and back. This doesn't mean that the performance by weiss and Dawson is worse than the one by weiss and Matthews. It's merely a *different* version. And it's this difference that influences the overall impression of the performance.

Dawson accompanies weiss and her vocal interpretation. The calm richness and depth of supporting basslines augment the deep calmness of weiss' delivery. The result is reminiscent of an audio book featuring background music. While it might sound less vivid than the version with Matthews, it is, however, a very typical Jazz & Poetry approach. Additionally, the differences between the two versions show weiss' flexibility in her Jazz & Poetry collaborations. weiss has no predetermined concept of rhythm or intonation. Instead, she creates every performance anew which highlights the importance of the moment of creative collaboration in her art in general and her Jazz & Poetry in particular. Collaboration has continued to be crucial to weiss, and in the last few years she exhibited a preference for collaborations with a jazz trio consisting of saxophone, bass, and percussion. An example of this can be found in the performance she gave on the occasion of her 90th birthday in San Francisco in 2018.

ruth weiss with Rent Romus, Doug O'Connor, and Hal Davis (2018)

A 90th birthday would be, for many, a day for looking back, for reflection. For ruth weiss, however, it proved to be a date for looking back graciously *and* ahead optimistically. Her Jazz & Poetry performance at the Beat Museum in San Francisco in 2018 attests to this; featuring old and new material, the performance included one poem of the *Desert Journal* cycle, a poem on the occasion of her 90th birthday, poems presenting her escape from Vienna in 1938, her life in Chicago in 1949 or in San Francisco 1952, and celebrating other "Women of the Beat". Altogether, weiss used this occasion to perform key biographical moments. The individuals in charge of this time travel soundtrack included saxophone player Rent Romus, bass player Doug O'Connor, and the percussionist Hal Davis – with whom weiss performed regularly. As she stated in an interview, weiss preferred collaborating with a trio of saxophone, bass, and percussion players because she felt most comfortable in this constellation, adding her voice as a fourth instrument, which a musician compared to a trumpet (Kluy, 88–9)[3]. Quartets with two horns, bass, and drums experienced their heyday in the 1950s and 60s with bands such as Gerry Mulligan, Ornette Coleman, and Eric Dolphy. Nevertheless, while the "ruth weiss quartet" didn't copy the sound of any of the aforementioned quartets, their name did echo the versatility and spontaneity of combo jazz of the 1950s and 60s. In what follows, I focus on the versions of "Bypass Linz" and part IV of the cycle "I Always Thought You Black" in order to analyze weiss' interaction with these musicians.

"Bypass Linz" is a collage of different passages of *Desert Journal* that ruth weiss assembled in 2008. Whereas these may have represented several strong

---

[3] The musician can be identified as Karl Schoen, one of weiss' saxophone players. In the documentary film *ruth weiss: One More Step West Is the Sea,* Schoen says, "ruth is such a strong, clear voice. One of the problems when you play abstractly, is it can get really amorphous. You're kind of all chasing each other around, but there's not that center there that you need to build on. And ruth delivers her poetry as a strong trumpet player would. She makes really clear, solid, heartfelt statements. And then that allows us to accompany her, kind of fill up around her." (Karl Schoen in *ruth weiss: One More Step West Is the Sea*, documentary film, dir. Thomas Antonic, 2021)

impressions in their original context, here they became a stream of associations connected to weiss' escape from Austria together with her parents. It comprised a collection of still not-forgotten but now only vaguely remembered fears and uncertainties. At the same time, the poem depicts the train trip from Vienna to Amsterdam (with an extended stop near Linz) with the expectation of "a ship / that took a trip / never to return" (weiss 2012, 35). weiss arrives in the USA and leaves behind her home, not with the feelings of loss and anxiety, but rather with mature courage: "little girl / leave your doll / she has told you / all she knows" (36).

As, in this case and unlike in the previous examples, weiss is now performing in front of an audience, it's interesting and worth considering her interaction with the listeners. Prior to the performance, an audience member asks weiss if she could improvise a poem or at least recite a poem that comes to mind spontaneously. She replies humorously but resolutely: "I don't memorize. It only appears where it feels like it." The audience laughs and applauds, and ruth weiss has once more established her approach to Jazz & Poetry performances. weiss repeats or skips single lines but never largely modifies her written poems in a live performance. Her focus in a performance is on the "how" and – even more importantly – the "when" of the words she speaks. Deciding which words she uses is part of the writing and not of the performance. However, as her response to the audience exemplifies, weiss doesn't rehearse together with the musicians beforehand, as she much prefers spontaneous reactions instead.

weiss starts this performance with the remark: "now close your eyes and just travel with ruth, ten-year-old ruth on a train ride, okay?" She foreshadows some information regarding the content of the following poem that also aims at guiding the manner of reception. The listeners are not to merely listen; they should imagine a train ride guided by the poet. weiss starts with the "Foreword" in a slow and calm voice. As the applauding audience interrupts the small pause between "Foreword" and "Bypass Linz", she takes up the role of a conductor calling: "Ready about to take off! Get on!" Then, weiss begins her recitation of the poem, in a voice that sounds somewhat threatening and forcing, changing the past tense of the first stanza into present tense to increase the immediacy of the story told and reduce the time that has passed. She quickly switches to an unagitated delivery, creating the impression of confidently talking about old memories that still move her. She invites the audience to "travel with ruth", which refers on the one hand to the autobiographical fearful ten-year-old ruth who escaped from Austria and on the other to the 90-year-old ruth weiss who is performing this escape artistically.

The trio supports this imagined journey. While weiss is presenting the "Foreword", Davis imitates the rhythm of a steam locomotive drumming off-beat over O'Connor's walking bass line while Romus adds the sound of a train whistle with a pipe. This imitation of a specific historical soundscape changes to a more autonomous sound when weiss begins reciting the poem. Bass player O'Connor is the driving force of the trio. While percussionist Davis either plays a minimalistic steady beat or dissolves the beat with drum rolls, O'Connor injects new impulses or creates the bridges between two parts. Romus' playing always interacts with both O'Connor and weiss. He adopts melodic or rhythmic ideas from O'Connor as well as thematic or expression-related ideas from weiss. For example, the trio begins with sustained sounds, accompanying the threat weiss is expressing with her voice in the first lines. Things then turn more melodic with respect to structure. weiss gives space after "so much warning / one was lulled by its constancy" (33) and Romus starts soloing over a short chord progression that doesn't really sound like a lullaby but more like "relaxed" jazz. When weiss re-starts, the musicians turn their melodic improvisation into a free-rhythmic, out-of-scale improvisation, little by little. After that, they return to calmer sounds, finally fading with phrases of short trills of saxophone and bass. The progress of these different parts perfectly fit the concurrent passages of the poem as this is a poem weiss and the musicians have performed before. The free musical part underlines the associations of the boat trip "on a sparkling rocking sea / darted with monsters" (35), and all musicians pause their playing on point when weiss recites "one is listening / for the sound that has stopped" (34). The end of the poem expresses a strong encouragement to leave behind the doll and the childhood it symbolizes in Austria. The trio illustrates this encouragement with a fading triller that does not rest on the tonic and could become the next starting point if needed. The musicians create a soundtrack of the imagined travel that sometimes sticks closely to the poem when they imitate the sound of a train, and in other parts creates a transposition of the textual meaning in music when they illustrate the encouragement of departure with a deceptive cadence.

The musicians choose a similar approach in the last piece of the set, part IV of "I Always Thought You Black". This poem perfectly works for a Jazz & Poetry performance because it portrays weiss as a young poet influenced by music. It takes us back to the life of ruth weiss in 1949, when she was living at the Art Circle in Chicago and starting her career as an artist: "my first home in bohemia. In the basement" (weiss 2011, 14). weiss lists the music she was listening to at that time, and portrays how she met Stuff Smith and Dizzy Gillespie, and where she wrote her poetry. She is amazed by Gillespie who is scatting "HOOPAKTECAH WHO PARKED THE CAR" (15) and who displays an artistic attitude towards life that she has made her own ever since. The poem emphasizes this in its last two lines: "walk home the 40 blocks or so. saves fare. forty some / years later am still riffing HOOPAKTECAH! —" (15) Despite the fact that weiss goes twice the distance in her performance than she does in the printed poem, which reads "walk home the 20 blocks or so" (15), and aside from a small slip of tongue concerning her rent, the poet doesn't deviate much from the written text.

The musical approach in this performance can be best explained with the passage where weiss lists the music she listened to: "I hear / BIRD / PREZ / LADY DAY / BUD / MONK / VILLA LOBOS / DJANGO / SARAH / SCHOENBERG / SHOSTAKOVITCH / BARTOK" (14). In the performance weiss pauses after "BIRD" for Romus to solo some fast Bird-like phrases, and also after "PREZ" (i.e., Lester Young) when Romus quickly changes to a more Prez-like melody line. Then, however, weiss drops names in a steady rhythm and the musicians proceed with their own music, including snippets of popular melodies and chord progressions. In doing so, they don't refer to every musician who appears in the poem. As such, instead of imitating a historical soundtrack, they are using the performance to create a musical version of the overall impression of the poem: weiss' life was surrounded by bebop, swing, and classical avant-garde music that helped her find her way as an artist. Mixing their own ideas with references to the musical tradition, the trio performs both a possible version of the vibrant musical scene that surrounded weiss in Chicago, at the same time that they claim their own position as part of a tradition that is still alive and very prolific.

In celebration of her 90th birthday, weiss performed poems that, taken together, tell her story of her journey to becoming a poet. As opposed to her poems from *Desert Journal*, which mainly consist of streams of associations connected by sound, weiss' later poems evoke the art scene of Chicago or San Francisco in the 1940s to 60s in a style closer to prose poetry. In this particular example, she performs these poems with a clear reference to the present, pointing out that the people she met in the past still influence her poetry and that they are, thus, still alive in her art. The musicians manage to transpose this precise sentiment through their music, referring to the tradition of both jazz and jazz combined with poetry. Thus, weiss and her musicians perform a living history, not only of ruth weiss' artistry, but also of the genre of Jazz & Poetry itself.

Conclusion

ruth weiss was one of the first to perform Jazz & Poetry in San Francisco in the 1950s and remained a prolific performer until her death in 2020. One reason for her continued artistic drive might have been her ongoing interest in creative collaborations. Poems like "I Always Thought You Black" introduce the names of artists that inspired her or that she worked collaborated with. In this light of heightened collaboration and artistic confluence, it is only natural that weiss would choose to perform this mutual influencing in Jazz & Poetry collaborations. Her recordings with Benfaral Matthews and Daney Dawson reveal how weiss performed with a spontaneity and flexibility that was appropriate to her musical partners. When she performed her tributes to artistic colleagues like in "I Always Thought You Black", she combined the memorized past with its ongoing influence in the present. This combination of past and present, which foreshadowed a contemporary form of Jazz & Poetry deeply rooted in tradition, could only have been achieved by someone who, like ruth weiss, understood the deep and complex connections of word and sound, of poetry and music. And someone like ruth weiss who, again, was there at the very beginning.

Works cited

von Ammon, Frieder. *Fülle des Lauts: Aufführung und Musik in der deutschsprachigen Lyrik seit 1945: Das Werk Ernst Jandls in seinen Kontexten.* Stuttgart: Metzler, 2018.

Antonic, Thomas. "Beat Authorship and Beat Influences in Austrian Literature." *The Routledge Handbook of International Beat Literature*, edited by A. Robert Lee, New York and London: Routledge, 2018a, pp. 157–170.

Antonic, Thomas. "'God's empty chair', or: You Can't Dig It – Jazz & Poetry of the Beat Generation and Beat-inspired Jazz & Poetry in Austria." *Jazz in Word: European (Non-) Fiction*, edited by Kirsten Krick-Aigner and Marc-Oliver Schuster, Würzburg: Königshausen & Neumann, 2018b, pp. 331–348.

Carden, Mary Paniccia. *Women Writers of the Beat Era: Autobiography and Intertextuality.* Charlottesville: U of Virginia Press, 2018.

Feinstein, Sascha. *Jazz Poetry: From the 1920s to the Present.* Westport / London: Greenwood Press, 1997.

Hrebeniak, Michael. "Jazz and the Beat Generation." *The Cambridge Companion to the Beats*, edited by Steven Belletto, Cambridge: Cambridge University Press, 2017, pp. 250–264.

Hunkemöller, Jürgen. "Jazz-Rezeption." *Die Musik in Geschichte und Gegenwart*, edited by Ludwig Finscher, Kassel and Stuttgart: Bärenreiter and Metzler, 1996, pp. 1421–1439.

Jackson, Travis A. "'Always New and Centuries Old': Jazz, Poetry, and Tradition as Creative Adaptation." *Uptown Conversation: The New Jazz Studies*, edited by Robert G. O'Meally, Brent Hayes Edwards and Farah Jasmine Griffin, New York: Columbia University Press, 2004, pp. 357–373.

Jones, Meta DuEwa (2011). *The Muse Is Music: Jazz Poetry from the Harlem Renaissance to Spoken Word.* Urbana / Chicago / Springfield: University of Illinois Press, 2011.

Kluy, Alexander. "'Just Keep Breathing!' Interview mit ruth weiss." *A Parallel Planet of People and Places: Stories and Poems*, Zirl: Edition BAES, 2012, pp. 87–90.

Schroeder, Tom, and Manfred Miller. "Ich bin seit Hellas ziemlich heruntergekommen…: Apropos Jazz & Lyrik." *That's Jazz: Der Sound des 20. Jahrhunderts*, edited by Klaus Wolbert, Frankfurt am Main: Zweitausendeins, 1997, pp. 643–652.

Weinreich, Regina. "Locating a Beat Aesthetic." *The Cambridge Companion to the Beats*, edited by Steven Belletto, Cambridge: Cambridge University Press, 2017, pp. 51–61.
weiss, ruth. ruth weiss performing with Benfaral Matthews, 1969. Unreleased reel-to reel recording, 1969.
weiss, ruth. *Desert Journal*. Boston: Good Gay Poets, 1977.
weiss, ruth. ruth weiss performing with Daney Dawson. Unreleased reel-to reel recording, Oct. 26, 1984.
weiss, ruth. *Can't Stop the Beat: The Life and Words of a Beat Poet*. Studio City, CA: Divine Arts, 2011.
weiss, ruth. *A Parallel Planet of People and Places: Stories and Poems*. Zirl: Edition BAES, 2012.
weiss, ruth. *Live at the Beat Museum 2018 and Other Recordings from the Archive*. Audio cassette. The Hague: Counter Culture Chronicles, 2019.
Whaley Jr., Preston. *Blows Like a Horn: Beat Writing, Jazz, Style, and Markets in the Transformation of U.S. Culture*. Cambridge, MA and London: Harvard University Press, 2004.
Wolf, Werner. *The Musicalization of Fiction: A Study in the Theory and History of Intermediality*. Amsterdam and Atlanta: Rodopi, 1999.

Peggy Pacini
"being tested": ruth weiss at the Summer of Love 2007

On September 2, 2007, over 150,000 people gathered at the Speedway Meadows in the San Francisco's Golden Gate Park to celebrate the Summer of Love 40th anniversary and revive the spirit and energy of the 1967 event and its prelude, A Gathering of the Tribes for a Human Be-In. The 2007 celebration hosted 1960s and more recent San Francisco rock and folk bands, indigenous tribe representatives, and poets, including Beat poets Michael McClure, Lenore Kandel and ruth weiss. Michael McClure and Lenore Kandel had also read forty years before, on January 14, 1967 at the Human Be-In, sharing the stage at the Polo Fields with Timothy Leary, 1960s Bay Area rock bands,[1] and Beat poets Allen Ginsberg, Gary Snyder, Lawrence Ferlinghetti. ruth weiss had not been one of those that read on that day back in 1967. She had just met her future longtime partner, the artist Paul Blake, at the Capri on Grant Avenue and they would soon head to Los Angeles as Summer of Love attendees swarmed onto the streets of San Francisco. However, by 1967, weiss had already long since become an active, established figure in the Bay Area art and poetry scene. Indeed, performing has been an old stomping ground of weiss ever since the mid-1950s when she had initially become involved in San Francisco's poetry and art community, pioneering in 1956 with saxophonists Brew Moore and Ben Webster Jazz & Poetry[2] reading sessions on Wednesday nights at The Cellar.[3] However, she often fails to be credited as "someone who paved the way" (Spandler, 228).[4] Drawing on an interview he conducted with weiss in 2014, Antonic provides some clues to that omission rec-

---
[1] This included Quick Silver Messenger Service, The Grateful Dead, Jefferson Airplane, Big Brother and the Holding Company and Blue Cheer.
[2] For a contextualization of Jazz and Poetry (see Antonic 183–186). In his article Antonic also recalls that weiss had already started reciting poetry in collaboration with jazz musicians in Chicago in 1949, and in New Orleans in the early 1950s.
[3] In his article dedicated to a history of the jazz scene in North Beach, "When Bebop filled the Night," published in *The Semaphore* (2011), Art Peterson rehabilitates ruth weiss' role in these sessions and cannot but admit that she has been the "unsung heroine of the Cellar." The article also offers a reminder of her grounded presence in North Beach at poetry readings by attaching the now well-known picture of weiss reading at the Grant Avenue Street Fair in North Beach, San Francisco in 1959 or 1960 (date differing according to sources).
[4] In fact, with the recorded version of their reading at The Cellar in 1957, Ferlinghetti and Kenneth Rexroth are instead often heralded as the innovators of poetry with jazz, as stated in Ralph J. Gleason's liner notes for the record.

ognizing that, in the 1950s and 1960s, she was "never part of an inner circle" nor ever actually wanted to be, and intentionally remained a writer in the background (182). Such omission also finds an explanation in the fact that, in the late 1950s and in the 1960s, female Beat poets were overshadowed by such Beat figures who were media-favorites like Ginsberg, Kerouac, Snyder, McClure or Ferlinghetti.[5] Categorizations, recognitions and labels aside,[6] weiss was a multitalented and many-faceted artist who had been experimenting in different areas of the arts ever since the 1950s, "a living example of the legacy of [the Beat] era and an incomparable and unforgettable experience when one has the chance to attend one of her live performances" (Spandler, 247). This chapter focuses on one of these experiences, namely, weiss' performance at the Summer of Love 40th Anniversary. It examines the nature of her performance within the combined framework of her performance practice and of the Beat poets' intimate relationship with the Bay Area bohemia and counterculture of the 1950s and 1960s. It also takes into consideration the Beats' central presence at the Human Be-In of January 1967, both as artistic and countercultural landmark, icon and filiation, and how the performance of their poems and chants partook in building up a communal event where the power of oral performance was advocated to explore new poetic conventions, foster community spirit, inspire social and produce physiological changes. The following analysis will thus shed light on three aspects. Firstly, the essence of weiss' poetry-composing and performing practice. In this respect, weiss' performance of three poems – "beast – be a saint," "speak for yourself," and "1967" – at the Summer of Love 2007 is unique and perhaps fitly exemplifies what lies at the core of her definition of "language" and is the essence of her poetics, which is precisely what this paper will discuss. A footage of her performance at the Summer of Love 2007[7] reveals something quite unexpected and challenging when compared with other live readings of the same poems,[8] and more generally with other poems read at other performances by the poet throughout her career. Secondly, her abil-

---

[5] See Estíbaliz Encarnación-Pinedo's chapter "Femmes: la Beat Generation [re]revisitée" in Penot-Lacassagne's *Beat Generation: l'inservitude volontaire*.
[6] For further insight, see Antonic 2015 on the canonization and marginalization process of Beat poets and within the circle of Beat poets, where he questions weiss' marginal(ized) place within the Beat Generation while offering a complete portrait of the poet, her many-faceted work and her recognition in time as a Beat poet.
[7] DVD without information about the filmmaker, private archive of ruth weiss, courtesy of Thomas Antonic.
[8] I will use here the more generic term of live poetry as identified by Julia Novak as "poetry performed or 'read' by the poet in front of a live audience" (21) in reference to Denise Levertov's essay on public poetry readings (1965).

ity as a poet-performer who had the ability to chisel her performance to the spirit of the era and place celebrated, lending it an essence of its own in the *hic and nunc* of its spontaneous production and building up a ritualistic communal experience. Thirdly, a deserved recognition given to the poet at the Summer of Love 40th Anniversary, altogether as a longtime West Coast poetry representative and as a female Beat icon.

Therefore, drawing on critical studies and methodology of live poetry (Levertov 1965; Novak 2011; Bernstein 1998), the emphasis will first be placed on contextualizing weiss' Summer of Love 2007 performance to delve into a performing practice that was altogether spontaneous and collective in nature. Then, the focus will be shifted to a compositional and performance technique that extensively relies on articulatory parameters (sound, vibration and resonance) and on "experiential meaning potential" (Kress and van Leeuwen 2001), i.e., the semantic impact of sound production. Eventually, the focus will turn to the social and communal function of this very performance, which becomes the area of dialogue where audience and poet-performer are invited to re-establish the communal bond and vision that this very celebration invoked.

By means of a comparative case study of the three poems performed at the Summer of Love in 2007 and of the same poems performed at the Porgy & Bess jazz club in Vienna, Austria, on October 21, 2013,[9] and recorded at Temple Studio in Albion, California, for weiss' CD *make waves* (2013), I examine the techniques weiss applies in performance to convey the colors, musicality, tone, rhythm and subsequently the meaning of the poems. For the most part, I will make use of a series of micro-analysis of what Bernstein calls "the audiotext" – i.e., "what the ear hears" or the audible acoustic text of the poem (12) foregrounded in the performance, its aurality – and of contextual factors of this performance (participants and spatio-temporal situation). This will contribute to comprehending how the Summer of Love 2007 performance released what weiss herself defined as the "free-flowing force moving outward from the unconscious towards self and other" (Grace 2004b, 58) that not only defines her poetic language but a poetics grounded in intermediality and performance.

---

9 An unreleased audio recording of this performance is archived on CD-R in the ruth weiss Papers at the Bancroft Library, University of California in Berkeley. A video clip of "1967" performed at this show can be found on YouTube. I will use both documents for my analysis.

"with the music inside/with the music outside:" making it happen in the here and now

The September 2, 2007 event had already started at eight o'clock in the morning, with Iroquois and Dakota Tribes chanting on stage. In the early afternoon, Country Joe MacDonald and his band had just left the stage when Paul "Lobster" Wells, the energetic MC, urged the audience to "welcome warmly please ruth weiss." With this laconic introduction, the poet arrived on stage, timid, fragile, and with a superb glittering pink butterfly covering the entire front of her blouse. She waved at the audience, moved to the microphone, grabbed it, launching into her four minute and 36 second performance of three poems: "beast – be a saint," "speak for yourself," and "1967,"[10] the latter of which was written particularly for this event. The poems aptly fit the celebration of the 1967 Summer of Love anniversary and echo the events that happened forty years before. Their titles are evocative of the visionary and esoteric experiences, creative energies and insights emerging in the late 1960s such as, to name but a few, the dawning of the Age of the Aquarius, the potential for a renewed civilization, the consciousness revolution and communal spirituality. Yet none of them were written in the era which inspired them, and were in fact composed between 1991 and 2007, "beast – be as saint" (1991), "speak for yourself" (1995) and "1967" (2007). The order weiss reads these poems in the Golden Gate Park 2007 differs from the arrangement of these poems in the book. Her reading sequence is carefully designed to follow a logical narrative which is calling for the spirit of 1967 to return.

Contrary to most of her poetry performances, there are no (jazz) musicians accompanying ruth weiss at the Summer of Love 2007, as there are, for instance, in Vienna in 2013, on the studio-recorded CD *make waves* and in general mostly when weiss is performing. On the contrary, in 2017 she is flanked by a ballet of technicians, impervious of her reading, moving to and fro in the background, busily helping the musicians of Canned Heat in setting up and tuning their in-

---

10 All three were published five years later in weiss' collection of poetry *a fool's journey* (2012). weiss had already performed "beast – be a saint" in that same year (2007). Various references are found for a concert entitled "Beast be a saint", which seems to be part of a tour weiss made in Austria in the fall of 2007, including one show at the MAERZ in Linz, on Sept. 26, announced as a music concert with ruth weiss (vocals-poet), Friedrich Legerer (saxophone), Martin Brunner (bass) and Gerry Krainer (drums), and another at the Porgy & Bess Jazz music club in Vienna, on Oct. 2., which announces ruth weiss (voice), Friedrich Legerer (saxophone), Gerhard Graml (bass), and Gerfried [a.k.a Gerry] Krainer (drums).

struments, as the band was to perform immediately after weiss finished. The setting was altogether overly different from small intimate venues (clubs and cafés) weiss mostly[11] read and performed her poems in, places where a close proximity to the audience also enabled those present to partake in the coziness and intimacy her performances generate. The open-air stage swarming with agitated technicians and a rock festival audience reshaped the reading and performing conditions the poet is accustomed to. If weiss is a poet-performer[12] that knows how to interact with and captivate an audience, one cannot agree more with Novak's conclusion that "a poetry performance is not a tangible object that can be studied independently of the spatio-temporal frame of its occurrence in any meaningful way" (174). The aforementioned stage setting and reading conditions that weiss had to comply with before the next set began constituted a definitely fortuitous situated-ness (Kress and Leeuwen 2001), one compelling weiss to offer a double performance, namely that of the poem and that of the performance of the poem. If this might appear as a challenge to weiss, whose body and voice at times exposed her nervousness, and who needed to adapt her voice to the particular soundscape situation and the acoustic surroundings of the venue, she proved to be a tightrope walker and a skillful craftswoman of words and riffs, always tailoring her flow to the balance and rhythm required. At the Golden Gate Park, weiss had no idea there would be a soundcheck during her performance and that she would have to compose with the brief structure of occasional riffs, and yet it confirms Spandler's statement that "no matter whom she perform[s] with, it [is] always important to her to see to it that music and voice me[e]t on equal terms" (230). Questioned by Nancy Grace on the ways in which her poetry and jazz intersect, weiss explained:

> My phrasing and rhythms depend on what I hear. It's a dialogue with the musicians. I never use music as a background. I give the musicians room to come up with riffs of their own. I lower my voice, raise my voice. I may repeat phrases. I may make up sounds. (…) It just comes to me. I don't know what I am going to do. And that's what is exciting (Grace 2004a, 66).

This cannot be more perfectly exemplified than in weiss' performance at the Summer of Love 2007, where she reacted to the bits and pieces of chord progres-

---

[11] Of course, weiss has also been invited to read and perform in institutional places, conventional auditoriums and festivals in the United States and abroad, but overall, it is obvious that the "performance space" (Novak 208) she prefers are small and intimate cafés and clubs.
[12] The term refers to the double role of the poet in live poetry pinned down by Novak whereby the poet has authored and performs his or her own poetry. For further insights into the role of the poet-performer in a live poetry performance, see Novak 175–194.

sion filtering in the background and played with them, naturally, as if it was her own band, constrained as she was to alter her delivery and reading of these three poems. Very soon into reading her first poem, her body started wriggling to the rhythm of her words and to the adjoining riffs and volume feedback. On the one hand, quite interestingly, rather than disturbing her, these seemed to pacify her as they offered the sort of dialogue with musicians she was used to when performing her poems. weiss started her performance with quite an authoritative vigorous tone, yet the volume feedback, though forcing her to stop, provided her a pause. From that moment on, weiss was back in her natural element when performing, playing with the presence of the musicians on stage and their occasional interference. On the other hand, something unpredictable was also occurring, as weiss was not only synchronizing and in tune with the musicians at the back of the stage, but indeed, musicians seemed to be so with weiss as well, most probably unwillingly.[13] The context of this performance thus turned out to be all the more challenging when, what might be interpreted as, or intended by organizers to be, a poetry interlude, turned very quickly into a genuine performance whereby weiss became poet-performer altogether, coping with the absence of her band of musicians, yet playing with those tuning in the background. This corroborates Grace's inference that weiss' method in public performance is altogether spontaneous and collaborative and contributes to creating a "choral voice" (Grace 2004b, 59). The choral voice Grace alludes to depends not only on the collaborative but also the collective action of the "participants," i.e. (poet-)performer, musicians, and audience[14] (Fine, 158). This is clearly perceptible in the three different performances under scrutiny, whose apparent differences result from a range of interrelated factors, times and location, and type of event.

---

13 Their responsiveness to her tuning to them is confirmed by weiss herself in an interview (weiss 2020).
14 In the fields of research on audiences and publics, different paradigms are described: the behaviorist, the Incorporation/Resistance (IRP), the Spectacle/Performance (Abercrombie and Longhurst, 1998), and the participation paradigm, which extends and completes the formers. In their essay on audiences (1998) Abercrombie and Longhurst look at how communication situations and modification of socio-technical environment affect the concept of audiences, which they conceive as a practice, and define as "groups of people before whom a performance of one kind or another takes place." According to them, the idea of performance, i.e., an "activity in which the person performing accentuates his or her behaviour under the scrutiny of others," is "critical to what it means to be a member of an audience" (40). They thus analyze a variety of communication situations involving a similar audience experience, to which the concept of performance is applied, to discuss the changes in the rules of interaction between the audience and the performance according to communication situations.

The audience at the Summer of Love 2007 was a festival audience, and a "simple audience" (44) according to Abercrombie and Longhurst's typology,[15] as was the audience at the Porgy and Bess jazz club in Vienna 6 years later. However, The Porgy and Bess is a medium-size Viennese jazz-club with space to accommodate an audience of 350 people, of which the larger part is comfortably seated, "which ma[kes] it easier to let the audience's attentive wonder" (Novak, 199). The audience's silence provided a listening atmosphere that favored some kind of interior experience on the audience's part, which was physically inclined to listen to the music of the poem, its tone, and even more so to its structure – a structure that, in weiss' case, was enriched and nurtured by the responses she received from the jazz band performing with her at the Porgy and Bess, for instance, when reading the poem "1967." On the contrary, the Summer of Love 2007 audience reacted and behaved differently within the frame of reference of a festival, with a majority standing, as tradition has it in open-air rock festivals, something that certainly adds a peculiar energy "as it require[s] a higher degree of wakefulness" (Novak, 199). The performance event space itself heightens the very role of the audience in this instance, as they are not "recipients" but rather participants, free to stroll about, talk during the performance and react to weiss' performance, which is why it almost appears as a rallying cry in terms of audience address and reception.

While in this case one might argue that during her performance, the audience's attention span might not be the best due to the soundcheck that is going behind weiss onstage, the reading is undoubtedly imbued with a ritualistic quality on different levels. weiss' performance acquires a quality that is reminiscent of the ritualistic and visionary nature of the countercultural events of the late 1960s, including the 1967 Summer of Love and the Human Be-In. In weiss' Summer of Love 2007 performance, a dialectical process occurred, whereby live poetry turned into a ritual and ritual was shaped by live poetry. On stage, the poet's and the speaker's voices merged to rally the addressed real-life audience – the participants in the ritual. The Summer of Love 2007 performance clearly assumed a ritualistic overtone within the hermeneutic frame of the poems. The three poems individually and as a group describe a ritualistic expe-

---

15 Abercrombie and Longhurst (1998) ground their theoretical framework in the integration of the Spectacle/Performance paradigm, which redefines what an audience is or what it does. Accordingly, they establish three categories of "the audience" – simple, mass and diffused. The "simple audience" is characterized by direct communication between audience and performer, localized/designated performance event place, high audience-invested ceremonial degree of the event, public performance, close involvement, high audience attention, and high audience-event distance (43–57).

rience, the steps of which are defined by each poem: a transformation, a test, a return.[16] The first, "beast – be a saint," is an exhortation to metamorphosis, a transfiguration, from beast into saint. The second, "speak for yourself," calls for an expansion of consciousness through thought-wave radiation (astrological influences), i.e., individual/personal waves that radiate out of our bodies and can reach others if attuned – that is, if you "relax / accept / take the energy," as the poem opening verses tell us. Moreover, the speaker of the poem also advises the audience to gain conscious control of their energy. Eventually, it is the very age of Aquarius that is "here to steer us / to a new dimension / to the next expansion." An Aquarian influence, as Elsa M. Glover emphasizes in her book *The Aquarian Age*, stimulates a will to break loose from tradition and authority, freely exercise one's own initiative and independence, develop one's own creativity, and stimulates self-awareness and embrace one's potential (9–10,13) (as suggested in "beast – be a saint").[17] In this instance, and to a greater degree than in other performances of the poem (i.e. in Vienna in 2013 and on the CD *make waves*), weiss mobilizes the audience's consciousness and sense of urgency, which she sustains throughout the performance by means of tone level, body language and rhythm, stimulating the audience into thinking about the message being delivered and the action suggested. The first two poems weiss reads are genuinely invitations to participate in a ceremonial liminal experience, whereby the essence of the addressee will be "tested" and their consciousness changed by means of a ritual, which the very circular, repetitive, anaphoric and chiasmic structure of the poem "speak for yourself" develops and which becomes a reality in the last stanza, as weiss' peaceful, slower tone in this last stanza suggests:

> relax
> accept
> take the energy
> and get a move on
>
> relax
> accept
> take the energy
> and get a move on

---

[16] In Victor Turner's essay on rites of passages *The Ritual Process: Structure and Anti-Structure* (1969), these phases would correspond to what he terms in the footsteps of van Gennep (1909): separation, transition, reintegration.
[17] On the Aquarian age, see also Roszak's essay *Unfinished Animal* (1975), especially the chapter "Aquarian Frontier" (19–43).

> being tested
> being tested
> being tested
>
> are we up to it
> are we into it
> *are we up to it*
> *are we into it*
>
> the AQUARIAN AGE is upon us
> *the AQUARIAN AGE is upon us*
> and our thought-waves most important
> make waves
> make waves
> make waves
>
> see *see* the picture you want to see
> see the picture in your dream
> see the picture & awake
> *see the picture & awake*
>
> AQUARIUS is here to steer us
> *AQUARIUS is here to steer us*
> to a new dimension
> to the next expansion
> *to the next expansion*
> *to the next expansion*
>
> being tested
> being tested
> being tested
>
> relax
> accept
> take the energy
> and get a move on
> (weiss 2012, 84–85)

The lines repeated in italics are not in the printed version of the poem but added by weiss at the Summer of Love 2007 performance. Therefore, this performed version emphasizes the ritualistic nature of the experience to an even greater degree as weiss moves the repetitive structure to the foreground by providing echoing patterns that act as call-and-response patterns. These call-and-response patterns, very common in weiss' performances, are central to her poetry, which is often rooted in the present. Verbs are declined in the present tense because weiss writes the poems, as she explains in an interview, "like I am there, so then the person who reads it, or hears it from me, is there with me" (Grace 2004a, 61). This has an impact, for instance, in the verb tenses she uses. As a case in point, "beast – be a saint," "speak for yourself" and "1967" are all written mostly either in the present tense or in the imperative, with natural return to the preterit to evoke 1967 remembrances, such as in the eponymous poem:

> 1967
> there was a dream
> 40 years ago
> SUMMER OF LOVE
>
> and we danced
> how we danced
> if it was the first time
> it was not the last time
> with the music inside
> with the music outside
> that was meant to last
>
> a shattering of glass
> the shards cut the heart
> wicked the reflection
> in the high noon-day sun
>
> a dream retreats
>
> there are dreams that do not vanish
> like the waters of the sea
> retreating to return a tidal wave
>
> BRING BACK THE DREAM
> (weiss 2012, 78)

The performance of the poems highlights this anchorage in the *hic and nunc* even further, making weiss' poetics – a threshold poetics flirting between the inside/inner and the outside/outer self and other – a poetics of consciousness, a poetics of the coming. Such a poetics, as exemplified with these poems in the Summer of Love 2007 performance, seems to rely extensively on weiss' use of the audiotext to release "experiential meaning potential." A micro-analysis of the first poem performed, "beast – be a saint," will exemplify what I call a poetics of the be-coming.

Performing a poetics of the be-coming: "beast – be a saint"

weiss opens the performance hailing the audience with a prolonged, amplified "hey" before she starts almost roaring the title of the first poem she is about to read, "beast – be a saint." Yet she spontaneously ceases, reducing the title to its first word, to the key concept of animality, which sounds slightly aggressive and bold. By immediately addressing the audience collectively and raucously as "beast," and rhythmically and acoustically playing with the vowel sound, stretching both diphthong <ey> (in "hey") and long vowel <ea> (in "beast") and ending these vibrations with a creaky voiced <a>, she immediately conveys a particular picture in the audience's mind. From the moment she grabs the microphone, weiss resorts to non-verbal sounds, amplified through the loudspeaker system, to catch the audience's attention and stir them, directly engaging them in both a close listening process of the performance and a total involvement in the performance experience. With such preamble to the poem, the first line, when uttered finally, becomes an existential predicate, mantra-like – "BEAST – BE A SAINT" – which she pronounces over four heavy beats. As a matter of fact, if originally the beast mentioned and summoned by the speaker might neither be the audience ("you") nor its members ("you"), weiss' performance at the Summer of Love 2007 reshapes the poem's deixis, as the beast[18] is invited to experience its own animality and holiness, not its bestiality. Each word, all monosyllabic, is granted equal importance and weight, though "saint" is pronounced with falling pitch. Her voice is animal-like, much like the feeling she wants to awaken in her performance, and yet, the roar fails to amplify and resonate outward, and it somehow resonates, much like an awakening cry, also reverberating inward, with rhythm and pitch modulations as particular vowels are assigned particular length and pitch, and with tension shifting from fast to slow tempo, conveying urgency and consciousness. The tone of her voice very much urges the audience ("it's time for transformation") to implicate themselves – "absorb this information / enter your potential / it's vital essential / your aura is calling / your calling can't stand stalling" – both individually – "your aura is calling" – and collectively – "we're on the move." But this is only possible if a collective will to perform it exists: "are we up to it / are we into it" ("speak for yourself").

---

[18] If contextualized with the period of the late 1960s, the beast conjured up can be read as the Beast of the Apocalypse, the AntiChrist, in the *Book of Revelation*.

From the very first line of the poem the speaker directly enters the fictive addressee's subconscious by performing a transmutation with the performative "BEAST – BE A SAINT!" The audiotext of "beast – be a saint" slightly departs from the published version, which reads as follows:

> BEAST – BE A SAINT!
> it's time for transformation
> absorb this information
> enter your potential
> it's vital essential
> your aura is calling
> your calling can't stand stalling
> we're all on the move
> to a better groove
> BEAST – BE A SAINT!
> BEAST – BE A SAINT!
> BEAST – BE A SAINT!
> (weiss 2012, 34)

In the printed version of the poem, the title is in lower case, while upper case is used in the core of the poem for the same line. The capitalized injunction is repeated four times, literally framing the poem at beginning and end. At the Summer of Love 2007, capitalization is transferred to another level of repetition. The line opens and closes the poem's performance, but the repetition of the ending line three times is transposed to a repetition of the word "beast," with the last line becoming "BEAST – BEAST BE A SAINT!", pausing after the second BEAST before concluding on falling tone and slowing down tempo, as if to actualize transfiguration. On the *make waves* recording, however, the pattern of repetition differs slightly, functioning as a mirror pattern providing a duplication of the two similar opening lines ("BEAST – BE A SAINT! / BEAST – BE A SAINT!"), with an alteration: "BEAST – BEAST BE A SAINT! BEAST – BEAST BE A SAINT!". Yet, on the release, "beast" is pronounced in much the same way as at the Summer of Love performance, with long vowels prolonged, but not aggressive, followed by "be a saint" falling tone (conclusive). The whole poem, in this recorded version, is delivered at a slight level pitch, rather playing occasional rising or falling pitch, which creates a syncopated pattern in the first two lines. Measures (bars) are displaced in comparison to the Summer of Love 2007 reading, which changes where the audience's attention is expected, something that in turn might alter the overall perception and interpretation of the poem. The binary rhythm of lines 4–6, however, is kept: "enter your potential / it's vital essential." The reading is accompanied by a jazz band that provides a musical dialogue with the poet and the message she delivers, which is also true for the 2013

Vienna live performance. The interaction one can infer influences weiss' performance as her voice is smooth, softer, more sensual altogether. The beast she seems to evoke in line 1 and title is uttered with amplification of the sibilant "s," playing perhaps more on the identification of the beast as the snake. Such amplification highly determines the nature of the beast she is conjuring in the audience's mind. At the Summer of Love 2007, weiss did not amplify the sibilant; her insistence was more on the vowels, which she stretched, and on the roaring of the sound produced, which rather evoked a wild beast.

Moreover, weiss also repeats line 5 twice, "your aura is calling," insisting on the call, while at the same time stumbling on "calling," so that the audience cannot hear distinctly "your aura can't stand stalling," and concludes the poem with a single "BEAST – BE A SAINT!" In addition, in the aforementioned performances of the poem, when ruth weiss is speaking the poem, the phrasing sounds quite different from what the page seems to suggest. In fact, when comparing the different performances under scrutiny of the poem, we realize that weiss doesn't pause at the same place. As a matter of fact, pauses tend to be displaced in her different readings, altering the perception of the line and of its meaning. If one considers the versions of the *make waves* recording and at the live performance in Vienna in October 2013, for instance, weiss pauses in the first line after the weak stress of the third syllables ("for"), therefore giving the preposition emphasis, while weakening the urgency ("time") of the transformation. In the *make waves* recording she repeats the same rhythm in the second line, and pauses after the third syllable, a weak one ("this"), here again shifting attention from the verb ("absorb") to the demonstrative determiner and hence on what "this" stands for, i.e., the information given in the following lines. These pauses seem to be responding to the jazz accompaniment in both performances. In the Summer of Love 2007 performance, pauses (|) divide the phrases as follows, with indication of stressed (S) and unstressed (U) syllables:

BEAST – BEAST BE A SAINT!	(S | SSUS|)
it's **time** | for **trans**for**ma**tion |	(US | USUSU |)
ab**sorb** | this **in**for**ma**tion |	(US | USUSU |)
enter | your po**ten**tial |	(SS | USU |)
it's **vi**tal | es**sen**tial |	(USU | USU |)
your **au**ra | is **cal**ling |	(USU | USU |)
your **au**ra is **cal**ling |	(USUUSU |)
your | **au**ra can't **stand stall**ing	(U | SUUSSU |)
we're **all** on the **move**	(USUUS |)
to a **be**tter **groove**	(UUSUS |)
BEAST – **BEAST BE A SAINT!**	(S | SSUS|)

weiss utilizes a balanced rhythm while also decelerating the pace for the central key line ("your aura is calling") to be heard, before fastening it up and accelerating pace, for the sake of urgency and imminence she is stating (lines 8–10). In the 1960s, the period the poem is referring to, such experience in one's own consciousness was tested through a variety of mind and consciousness altering drugs and practices such as LSD consumption or Tantric Yoga, among others. The transformation suggested by the opening line and title itself is conveyed by the very structure and construction of the line itself. The following microanalysis attempts to provide a clarification as to how the saintly nature called upon is already contained in the bestiality suggested and which, once deconstructed and transformed, allows for saintliness to become incarnate.

A closer look at the printed version of the poem "beast – be a saint" provides insights as to how weiss composed the poem. It also offers a guideline to her live readings of the poem and an understanding of how she improvises in performance from and with this compositional technique. The printed version of the poem reveals a simple and concise compositional frame, both formally and rhythmically. The first line is the general framework from which the structure is going to develop and expand by a series of line, word and sound repetitions, alterations, syllabic variations, and extensions largely dependent on rhythmic patterns or cycles. On the one hand, the poem is framed by the eponymous title line "BEAST – BE A SAINT!" in upper case, the structure of which consists of three accented monosyllabic words, the first, "BEAST," being developed in the three monosyllabic words of the second part of the line, "BE A SAINT," graphically cut by three dashes "—." On the other hand, it is framed by an alliteration in *b* and a consonance in *t*, somehow providing the word "beast" to morph into "**be a saint**," with the first alliterating word gradually transforming into the second alliterating word "be" and morphing into its final entity "saint." The subject is morphing into the object, by a gradual process of dismemberment of the subject ("beast") into the targeted object ("a saint"). The printed poem can only offer punctuation for this chorus: three dashes separating "beast" from "a saint," but these dashes could also be interpreted as the opposite, as elements providing a connection. Indeed, when performing this line in the live performances under scrutiny, weiss stretches utterance of the word "beast," then pauses before she utters in a single breath falling tone "be a saint," thus sealing transmutation and operating the transfiguration through the performative act. In addition, the three dashes might be read as a deforming mirror, allowing words to be transformed, morphed into other words. Equally, they could rhythmically also be interpreted as a pause, announcing the line's 3/4 beat, with each dash acting as the graphic representation of both the line beats and the words "be a saint," which, when fused, produces "beast." The dashes signal the transformation announced in line 2, and at the same time they provide a method of reading and performing the poem as a musical score, alternating 3/4 beat lines with 2/4 beat lines, and closing the poem with a triple repetition of the mantra-like "BEAST – BE A SAINT." This compositional method, which is deeply rooted in weiss' intimate relationship with jazz, also provides an adequate score for weiss to freely improvise from such a framework in a performance (upper case words, punctuation, rhythmic and stylistic repetitions). As she clarified in an interview with Nancy Grace: "for the thing to really work the meaning and the reverberation of the sound have to hit at the same point" (Grace 2004a, 65). For weiss, nuances are a form of being tuned in, as she is someone who has an ear for phrasing and rhythms and who extensively uses the memory of sound, which on the written page partakes in weaving a semantic connection through repetitiveness. It is something she derives from jazz improvisation and the very essence of which, as Christian Béthune clarifies, is extensively the product of a mimetic device founded on memory, and not only that of the audience, but of the performer as well (452–453).

In this sense, weiss' poems are expertly composed scores wherefrom, in performance, she rhythmically nuances and durationally varies motifs. In fact, comparing the performance of these poems at the Summer of Love 2007 with their performance in Vienna in 2013, one cannot but notice how, quite naturally, weiss uses her own musical acuity and sound memory to engage her audience into their own memory and interpretation of the words of the poems in relation to their own experience and/or vision of the late 1960s. In performance, this sound memory is also a crucial aspect of the participants' interaction and is very much conditioned by the performance space, and the experience awakened, i.e., an invitation to personally and/or collectively experience the message of these poems. This is largely perceptible at the Summer of Love 2007, for instance, in the performance of this opening poem, "beast – be a saint," where the repetitiveness and sound memory that weiss creates aids in developing a communal experience.

As will be considered in the final part of this essay, I firstly examine how weiss manages to create this communal experience, and secondly how, by doing so, she seems to perfectly bring the awakener, the prophet-like Beat poet to life – as was personified in 1967 at the Human Be-In, mostly by Ginsberg, but also by McClure, Snyder and even Kandel.

"all the pieces fit singing to each other" (weiss 2011, 30): conjuring up the spirit of 1967

What marks a live poetry performance, as Novak pinpoints, is the fact that its production and reception is simultaneous and collective. Consequently, it mostly relies on the "common effort of poet and audience to make it happen in the here and now" (173) and on performer-audience relations, i.e., how they influence and interact with each other.

As noted earlier, weiss' poems function structurally on the basis of repetitive phrases, both at the composition and the performance level. The ritual nature weiss conjures up in the performance at the Summer of Love 2007, as suggested before, can be examined by means of an analysis of the call-and-response pattern in these poems. On the written page, weiss already provides a call-and-response pattern, whereby the speaker responds to himself or herself. However, when she performs with a jazz band, the antiphonal pattern is clearly redistributed between herself, poet-performer (soloist), and the band accompanying her. I have already demonstrated earlier how the Canned Heat musicians rehearsing on stage during weiss' performance played a part in reactivating this pattern. If such is the case, might this also have contributed to the performance gradually turning into a communal moment? Undoubtedly, what happens on stage and in the audience determines quite differently and significantly how ruth weiss utilizes the poem as a medium of communication, performance and, individual or group experience. At Golden Gate Park, weiss needed to ensure that the audience was with her. Consequently, the sense of direct address to the audience that was provided by her greeting "hey beast," opening the performance, was heightened fivefold in this instance in that the poet vehemently implicated the audience on hand in the process and stages of transformation. weiss' call, however, required some time for an active response from the audience. She started delivering her first poem "beast – be a saint" at a relatively high volume, increasing the volume on particular words that provided semantic information (*time, absorb* and *enter*) or emotional force (*vital, essential, better*), definitely conveying an authoritative tone and demanding active response. This was intended to allow the audience to be directly engaged with these imperatives ("**absorb** this information / **enter** your potential") and realize the call as an imperative ("it's vital / essential"); that said, the audience failed to contribute actively to the pattern. Response, instead, was transferred onto another level – as suggested in the first part – to that of the "musician-performer communication," (Elam, 86) to borrow and adapt Keir Elam's critical tools for analyzing the audience's response to the activity of the poet-performer in a theater context. This pattern of "spectator-performer," which includes the audience in the context of a public reading or performance, remained monovocal at the outset and was hardly a participative exercise at the Summer of Love 2007. weiss offered both call and response. However, the audience became more involved in the call-and-response pattern produced in the second poem, "speak for yourself." In fact, the audience started cheering when weiss asked: "are we into it / are we up to it." This feedback thus invited weiss in turn to repeat these two questions, which again received approval from the audience. weiss then launched into the next line, "the AQUARIAN age is upon us," and the audience joined in repeating "the Aquarian Age," which yet again incited weiss to repeat her line twice. Now the audience was with her. A member of the audience eventually took up the call-and-response pattern by repeating the entire last line of the last poem "1967," shouting "in the high noon day sun." What Malte Möhrmann interprets as a "*manifest*" audience response (Novak, 196) – which includes extreme audience reactions such as cheering, whooping, but also phrases or line repetition – provided the sense of dialogue weiss was used to with her musicians, freeing her to improvise and spontaneously repeat or alter lines. The repeated stanza "a shattering of glass / the shards the shards cut the heart / wicked the reflection / in the high noon day sun" ("1967") was performed with an omission of the adjective "high" in the last repeated line. Undeniably, one realizes how both weiss and audience are here partaking in the cyclical construction of the ritualistic pattern of the performance generated by both manifest audience response and what Erika Fischer-Lichte refers to as a self-regulating "*feedback loop*"[19] on the part of the poet-performer. weiss' use of repetition was a sign that she was reacting to the audience's signals she was receiving. In the last poem, "1967," weiss, however, had to impose her voice and her rhythm as MCs and musicians had begun to move to and fro once again; she now embodied another role as she tried to regulate and orchestrate both the audience and the musicians with her hand, resorting to a conducting gesture she often used in performances.

It is however the performance as a whole of the three poems in the context of the Summer of Love that powerfully gives a special contextualized meaning to the performance as it seeks to convey the spirit of "1967," the eponymous poem weiss closed her performance with. By reclaiming such a spirit, the performance conjured up a collision, both temporal and spiritual, by deeply rooting it in the *hic and nunc* of the performance act. What is interesting in this perfor-

---

[19] Term coined by Erika Fischer-Lichte in *Ästhetik des Performativen* (2004) (qtd. in Novak, 196).

mance is that these three poems, read in that specific order ("beast – be a saint," "speak for yourself," "1967"), can definitely be heard as one poem, with different transformative sections, each poem guided by a special request: "BE A SAINT," "relax / accept / get the energy," and "BRING BACK THE DREAM." Firstly, as noted before, the titles of these three poems are delivered in such a way – fluidly merging with the flow of delivery – that the audience might simply have interpreted them as lines in the one poem. This is all the more true for the first two poems performed, with just a very slight pause between the two, drawing the audience into thinking it to be one and the same poem. This is also reinforced by the fact that weiss did not pause to take another sheet from her folder. When it comes to the poem "1967," weiss changed sheets, showing them to the audience, and before she started with this last poem, the band on stage engaged in a few louder riffs. The audience was whooping, and weiss refocused their attention with a "can you hear me." The question here has less to do with whether they can hear her properly than it does with whether, as asked in "speak for yourself," they are "into it," i.e. in tune with her, experiencing the poems. So when weiss resumed with "1967," the first line, reduced to that very date, was remodeled into a signifier meant to stir up memories in the audience and remind them of what the actual first line of the poem, here repeated in the performance twice, sought to "bring back" in the first stanza: "there was a dream / there was a dream / forty years ago / summer of love."

The very singularity of weiss' performance at the Summer of Love 2007 was a tempting case study to delve deeper in her craft both as poet and performer, one she had been perfecting ever since her very first poetry readings at The Cellar in 1956. Shedding light on her compositional practice, these poems performed at the Summer of Love 40th Anniversary, with all the performance space difficulties she was confronted with, in addition to the fact that she was reading unaccompanied by a band as she would have been used to doing, confirms her natural inclination, but most of all her drive, to orchestrate her poems as scores to be performed. As such, the Summer of Love 2007 performance is quite a singular one, though weiss seems to rely on a set of typical reading habits, as the challenging surrounding aspects of the performance compelled her not only to integrate paralanguage elements (non-verbal vocal features) to her reading but also to heighten communicative dimension with the audience, which often altered how the poems were interpreted. What is clear is that the varying natures of these performances of these three poems in Vienna in 2013, on the CD *make waves* and most of all at the Summer of Love 2007, instill the written text with a will of its own when filtered through the oral performance at a specific time and in a particular space, turning it into an unstable medium allowing time and space to penetrate and modify reception and, perhaps even more so, perceptions of the poem. This is decidedly clear at the Summer of Love and might come as a counterargument to the often-cited argument supported by those who, like David Groff (2005), oppose literacy to verbal performance, and consider that the value of a poem can be lessened by that performance. Arguably, in weiss' case, the poem in performance is imparted with a life of its own, often imbued with a communal dimension when performed and appropriated by the audience at particular events in distinctive spaces and times. Ultimately, the singularity and ritualistic nature of her performance at the Summer of Love 2007 certainly confers weiss a guru-like, prophet-like status, often denied to female Beat poets and easily bestowed on male Beat poets. Yet if she never meant to be considered either a prophet/guru or the 'Beat Goddess,' a moniker *San Francisco Chronicle* columnist Herb Caen dubbed weiss with in 1993, it cannot be denied that her voice is the vehicle of a poetic magic that the performance of her poems only heightens and radiates.

Despite the fact that it took a long time for academia and critics to delve deeper into weiss' work, and as Antonic warns, given that it is clear that one should be cautious with assigning labels or creating categorizations, weiss has always found her crowd among small and larger audiences and has not garnered any kind of media recognition, or so it seems. When, at the end of her performance at the Summer of Love 2007, the MC came back on stage to a cheering audience, he rightly lauded and celebrated "ruth weiss, one of the original Beat poets, ladies and gentlemen." If the adjective chosen to qualify weiss was perhaps meant to accentuate the Beat filiation at the Celebration and convey authenticity, it definitely lends credit to weiss and her role in the history of the Beat Generation while reintroducing her in the context of a literary and performance history that largely failed to celebrate her true singularity as a poet-performer who lives, writes and performed...to the beat.

Works Cited

Abercrombie, Nicholas and Brian Longhurst. *Audiences: A Social theory of Performance and Imagination.* London: Sage, 1998.

Antonic, Thomas. "From the Margin of the Margin to the 'Goddess of the Beat Generation'. ruth weiss in the Beat Field, Or: 'It's Called Marketing, Baby'." *Out of the Shadows. Beat Women Are Not Beaten Women.* Ed. Frida Forsgren and Michael J. Prince. Kristiansand: Portal 2015, pp. 179–199.

Bernstein, Charles. *Close Listening: Poetry and the Performed Word.* Oxford: Oxford University Press, 1998.

Béthune, Christian. "Le jazz comme oralité seconde." *Revue française d'anthropologie*, 2004, pp. 443–457.

Cohen, Allen, ed.. *The San Francisco Oracle*. Facsimile Edition. Berkeley: Regent Press, 1991.
Elam, Keir. *The Semiotics of Theater and Drama*. London: Routledge, 2002.
Encarnación-Pinedo, Estíbaliz. "Femmes: la Beat Generation (re)visitée." *Beat Generation: l'inservitude volontaire*. Ed. Olivier Penot-Lacassagne. CNRS Editions, 2018. 143–153, 2018.
Fine, Elizabeth C. *The Folklore Text: From the Performance to Print*. Bloomington: Indiana: Indiana UP, 1984.
Glover, Elsa M. *The Aquarian Age*. [Morrisville, NC]: Lulu.com [Lulu Press], 2013.
Grace, Nancy M. "Single Out." *Breaking the Rule of Cool: Interviewing and Reading Women Beat Writers*. Ed. Nancy M. Grace and Ronna C. Johnson. Jackson: UP of Mississippi 2004, pp. 55–80. [2004a]
Grace, Nancy M. "ruth weiss's *Desert Journal:* A Modern-Beat-Pomo Performance." *Reconstructing the Beats*. Ed. Jenny Skerl. New York: Palgrave Macmillan, 2004, pp. 57–71. [2004b]
Groff, David. "The Peril of the Poetry Reading: The Page Versus the Performance." 26 January 2005. Web: https://poets.org/text/peril-poetry-reading-page-versus-performance. Accessed 1 June 2020.
Knight, Brenda. *Women of the Beat Generation*. Berkeley: Conari , 1996.
Kress, Gunther, and Theo van Leeuwen. *Multimodal Discourse: The Modes ad Media of Contemporary Communication*. London: Hodder Arnold, 2001.
Levertov, Denise. "An Approach to Public Poetry Listenings." *The Virginia Quarterly Review*, Summer 1965, pp. 422–433.
Novak, Julia. *Live Poetry: An Integrated Approach to Poetry in Performance*. New York/Amsterdam: ,2011.
Peterson, Art. "When Bebop Filled the Night." *The Semaphore*, vol 7, no. 1, Spring 2011. Web: https://d6dddace-3783–4032–909b-f74b5a41095d.filesusr.com/ugd/38a738_672a4233940842deb1c41f9b6d43b848.pdf. Accessed 1 March 2020.
Roszak, Theodore. *Unfinished Animal. The Aquarian Frontier and the Evolution of Consciousness*. London: Faber and Faber, 1975.
Spandler, Horst. "ruth weiss and the American Beat Movement of the '50s and '60s." *No Dancing Aloud / Lautes Tanzen nicht erlaubt*. By ruth weiss. Bilingual Engl. / Germ. Germ. transl. Horst Spandler. Vienna: Edition Exil, 2006, pp 213–247.
weiss, ruth. "1967." Vienna, 2013. Video. Web: https://youtu.be/ZIeGrsDSkJ8 Accessed 1 March.
weiss, ruth. *Can't Stop the Beat. The Life and Words of a Beat Poet*. Studio City [Los Angeles]: Divine Arts, 2011.
weiss, ruth. *A Fool's Journey/Die Reise des Narren*. Bilingual English/German. Trans. Peter Ahorner and Eva Auterieth. Vienna: Edition Exil, 2012.
weiss, ruth. *Make Waves*. Vienna: Edition Exil, 2013. Audio CD.
weiss, ruth. Personal interview by Tate Swindell, 15 Jan. 2020, San Francisco. Unpublished audio recording (courtesy of Thomas Antonic).
Whaley, Preston Jr.. *Blows Like a Horn. Beat Writing, Jazz Style, and Markets in the Transformation of U.S. Culture*. Cambridge, MA: Harvard University Press, 2004.

Caroline Crawford
Oral History Interview with ruth weiss

The interview was held in Albion, California, where ruth weiss lived, on July 11, 2019.

Crawford: I'd like to start with your poetry that I would describe as intimate, unleashed, personal. What would you say?

weiss: That unleashed, I like that, that's wonderful. People ask me what style I do, and I say, "I have the poem tell me. I don't know what style it is until it gets started," and I say, "when I, for example, do a poem about a person, a place, I let that theme, that person who has part of their life in my poem, enter it." So I don't know how it's going to go, and my whole theme, my whole philosophy is improvisation. That's my whole thing. It's complete improvisation. I never know what, you know. I do haiku as a form, and I have done a few sonnets, but I'm really mostly into improvisation.

Crawford: I wonder if you think this is accurate: Ferlinghetti said in *A Coney Island of the Mind*, he said, "Your poetry is told in flashes, in arias." Arias would mean operatic, that it's operatic. On a grand scale!

weiss: Ah. Mm, well, he's Italian.

Crawford: [laughs] That's a great answer. Let's go to Berlin, your birthplace. What do you know of your birth? In *Can't Stop the Beat*, your book of poems, you write: "I had just been born. The nurse asked Mutti if Papa was black. It was the 20s and jazz musicians were made most welcome in Europe, 'What do you mean, let me see her...this is my baby she looks just like Oskar,' Mutti cried." (weiss 2011, 9)

weiss: I'm Austrian because my father was, but I was born in Berlin, Germany. The reason was my father was a journalist, and he would travel all over, and at this particular time, we were so-called stationed, in Berlin. That's where my father was at that time, and so, I was actually born in Berlin, and we lived as just the three of us, my father, mother, and I. I do not have siblings. We lived there until I was five in 1933, and that's when Hitler took over Germany. So, we fled. We didn't flee; at that time, we just took a train and went to Vienna, back to my father's mother's home, who was my grandmother in Vienna, and we lived there from 1933 to 1938, when Hitler followed us. What do I remember about my first five years? Well, one thing: I wrote my first poem when I was five. I could tell you the poem. I'll

| | read the German one, or say it, and then I'll translate it, and it was funny: in a very recent book, we finally published my first poem.
Now, all right, I'll tell you the poem. "es war einmal ein bär / der hatte braune augen / spazierte hin & her / und wollt zu gar nichts taugen." (weiss 2012, 8) "once upon a time a bear / had brown eyes / ambled back and forth /and did not want to be of any use,"[1] and then as I say, that was the first beatnik that I – [laughter]
In Vienna grandmother had a big apartment that covered the whole floor with about seven rooms, and she turned it into a rooming house for students. The medical university was just a few blocks away, so students would rent it for a whole semester, and then stay at my grandmother's. |
Crawford:	You mention Rumanians, Australians, at least one Hindu, Japanese and Chinese, and you slept in the bathtub when the rooms were all taken (in weiss 1978, n.p.). [laughter]
weiss:	Yes. And I went to school. Now, we're talking about two different places. There are even pictures of me with friends that age – I'm talking about kindergarten age – and they're all very German, and they were part of the people I knew, and overnight, they couldn't talk to me, because of Hitler. Because I was Jewish.
Crawford:	Because your family was Jewish. Well, when you went to Vienna, did your parents tell you why? What did you think was happening?
weiss:	Well I knew something was wrong, because they had very worried looks on their faces.
Crawford:	And so when did things start to get more serious?
weiss:	In Vienna. My mother and I were walking near the temple, and the Torah, which is the five books of Moses, in the scroll form – they were attacking the synagogue, and there weren't any people in it. I don't know if they torched it or whatever they did, and they threw the Torah into the street, and my mother went, walked over to pick it up, and there was a cop who used to help us across, and he knew my mother.
She went to pick up the Torah, and he came over and he – "Don't, Frau Weiss, don't, your life," and she just grabbed my hand and we walked away and never looked back, but it was dangerous if she had picked it up. I think that's in my book, *Single Out*. My refugee story is in that part.

---

[1] Translated slightly differently in weiss 2012, 8.

	My best girlfriend at that time, who I went to school with for four years, and I were walking right down the street from where she was living. It was kind of a park, not a real big one. The whole point is that my girlfriend Susi and I, and here we are ten years old, so we're not little kids and we know what's going on, and there are big signs, on the park, "*Juden Verboten* – Jews Not Allowed." And, just then about twenty-five young men in brown shirts and swastikas showed up at this park, and they were like maybe a block away. They were right there and we were here, and they were coming this way. They didn't say anything. They were just walking, and my girlfriend, she said, "I'm proud to be a Jew," and I thought, oh God, this is it, right? And they stood there, and they started to laugh, and laugh, and then they walked away. They saw these two little girls saying, you know, and they laughed. Of course I wrote, "We never told anyone. It was foolish." (weiss 1978, n.p.)
Crawford:	How did you get out of Vienna and come to the States? You went through the Netherlands, I think.
weiss:	It's a long story. All right. Let's get started. So, different students would stay in our rooming house, and usually, people rent by the month, not by the day, and most of them were students, young people in their late teens or early twenties, and one day this man appeared who was my father's age, and they became good friends, from America, who lived in New York. There's a section that's very Jewish. It's Coney Island, and then there is another place, Brighton Beach, that's where I did most of my swimming. Anyway, he came, and he studied some at the university which is near there, and he and my father became good friends. So, when it's all messed up here with Hitler, my father immediately got in touch with him to help us get a visa. Now, he didn't have much money, but people in the congregation did, and he was able to find somebody who sponsored us. So, now Hitler has annexed Austria, where he was very welcome at the time, and on the *Kristallnacht*, which was the night that they broke windows after windows of any Jewish store – it's called the *Kristallnacht*, and oh, what that was about: a Jew in Paris had killed a Nazi. This was a revenge.
Crawford:	Retaliation.
weiss:	Absolutely. And so, that night, the Nazis, SS, or whoever they were, the Black Boots, they had the name of all the Jewish men, and that night, they came to every Jewish home in Vienna and took the men

– they didn't take the women and children yet – so my father was one of the people picked up.

Within two weeks, he walked in, and before we had talked tender words he said, "Pack immediately. We're leaving tonight." We didn't ask any questions. We just packed, just a little, and just packed some warm clothes. Now this was November of '38, and the visa hadn't arrived yet, okay? I know my parents spent many hours in lines getting those papers.

They didn't say he could leave; they just released him. Now we really realized fully what's going on, in those two weeks he was in the jail, and he said, "We're leaving tonight," and just, "we're leaving in a few hours. There's a train going to" – I believe it was Innsbruck, and we were going to Holland through Switzerland.

So, we packed, and left; took a train to Innsbruck, tried different ways to get across the border – I mean, we tried, and it was pouring, and pouring, and pouring. I've never seen such weather in my life, and here we are with a guide.

Then we tried a boat, then the boat capsized, and all kinds of things happened, and so here we've spent all our money. Here we are in Innsbruck, sitting in the railroad station, nobody around. Any minute, they're going to come checking out, like, we had nothing to lose but we didn't know what to do, and a woman comes by. She doesn't look at us; she just says, "Follow me."

Well, we had nothing to lose. We were going to get picked up. They're going to see us. We were obviously refugees, and we followed her into the house – she lived right near the railroad station – and she said, "What are you trying to do?" We said, "Go back to Vienna." She said, "Well let me see what I can do. Now get some sleep," and she gave us this bedroom, and then my parents went to bed, and here was the couch for me, and I have always been an insomniac.

I heard voices, and I got up and peeked out, and saw this man sitting and talking to this woman, and I didn't want them to see me, so I just backed off and I went back to sleep. She woke us up, made breakfast for us. We get to the station, and now it's busy. Everybody's going to work, and it's a full working train, and she says, "Here are your tickets. Get on this train number so and so. Don't talk to anybody; say nothing. Just go right on that train, and this will take you all the way to Vienna. You don't have to change trains or anything."

	So then, we left her house, and crossed the street, and went to look for that train number, and there're lots of people now. They're all going to work or whatever, right, business as usual, and I hear this sound – I'll never forget it – the boots striking, a boot, and I looked up to see who this was, and I couldn't believe what I saw, and I turned away very quickly. It was the man in the apartment who got us the tickets back to Vienna. And now, we got on that train. We get back to Vienna, and walked back to the house with my grandmother there, who didn't know –
Crawford:	She wasn't going with you.
weiss:	No. She was too old. We just had the visa for the three of us anyway, but she wouldn't leave her house. Yes, this is something, when I saw how fantastic it was that we were stopped from going on to Switzerland. See, the sharpshooters in Switzerland were shooting over our heads. They didn't want to kill us. They just let us know that there's no way of coming in, and we found out much later that the borders were really, really absolutely closed, and since then, one of my philosophies is, trust the bend in the road. I've noticed that almost everything in my life where something really negative happened, it was in the long run the best thing that happened. If we had been able to go into Switzerland, escape, we would have spent the war not in a concentration camp, but we couldn't go anywhere; we couldn't have gone anywhere else. We would have been in a camp, but not a concentration camp, but I've heard stories that it was not very pleasant. In fact, one of the rules that the Swiss had which I thought was horrible – I mean, they didn't hurt you, they gave you the food – they didn't allow any of them to play music.
Crawford:	The Swiss?
weiss:	The Swiss kept their temporary prisoners – as I say, they fed them, they were not beaten, but can you imagine have a Gypsy, a Roma, being forbidden to play music? That's kept them alive.
Crawford:	Did you lose other relatives too?
weiss:	All of them. My cousin's same age as me that I spent a lot of summers in Yugoslavia with her, was the youngest. She was exactly my age, and we used to watch Shirley Temple films together.
Crawford:	But you made it, and eventually settled in Harlem. Why Harlem?
weiss:	Yes, that was strange, there. All right, what happened? They put me into a children's home, a Jewish children's home in the Bronx. Nobody spoke German there, by the way, or Yiddish even, and it was

a very nice, clean, et cetera, children's home, and I was enrolled in a public school, except I didn't know a word of English, so I'd sit in the back of the class and make drawings of the students, and give them those as a way of connecting.

Well, in order to be in that particular place, you had to be under eight. Well they finally found out that I was over the age that they took children, so, they shipped me over to Harlem to a section that used to be Jewish, and that's why, and that was a Jewish children's home in the middle of Harlem, and I went to school in that neighborhood, and of course since it was Harlem, everybody was black, and they were so wonderful to me, the kids.

Crawford: They were accepting.

weiss: Oh, more than accepting. They wanted to know where my mother and father were, and I said, "Yes, they visit me on Sundays," and they said, "Where are your brothers and where are your sisters?" and I said, "Well I don't have any" – "You don't have any?" – and they couldn't understand that. They've never heard of a family without children.

They weren't too insistent on me being back at the home on time, as long as I was back in time for supper, so afterwards, they thought I was very strange, so they took me to their families and I would have supper with them. I'd be late and they were upset. I said, "I'm sorry." "Well you can't have supper tonight." I said, "Well that's all right." [laughs] I had very good suppers. And, slowly, I actually started learning some English. I remember a particular night I woke up, I was dreaming in English, my goodness!

Crawford: Then you know that you've really learned it –

weiss: Yes, right. But meanwhile, before I got moved over there and I was still in the Bronx, I met a girl who spoke German. Her parents spoke German, and they were Germans, but she was somebody else I made friends with, and I would draw them pictures from actually German kinds of fairy tales, about *Nibelungen.*

Crawford: As part of your schooling?

weiss: Well, I would just in the class make drawings, of these tales, mostly *Nibelungen.* And, I remember doing a lot of those drawings, and the teacher seeing those, enrolled me in an art class, and I've always thought if I had not left that one and stayed more connected with the school that I was going to at that time, if I would have grown up in New York, and maybe been part of that art scene, you know

	what I mean? But they moved me to Chicago. Oh, Chicago is another story.
Crawford:	Well, you said in your poetry that you felt later on in your life that maybe you hadn't put yourself forward strongly enough, you didn't record as much as a lot of other people, and you said, the reason for that was that you had to hide a lot when you were a little girl. Is that true? You were shy.
weiss:	Yes, and the hiding was probably a lot about my life, that I learned to be quiet about things in my life, so –
Crawford:	You mean you didn't talk to your parents much about that?
weiss:	Oh, I talked very little to my parents. First of all, they were both working.
Crawford:	What was your mother doing?
weiss:	She worked in sweatshops.
Crawford:	Doing what, sewing?
weiss:	Yes. She was always the one who had the most done. She was very oriented into – oh, here is part of the hiding that I just realized. It was my mother who wanted to hide me. We were in the Automat eating sandwiches. And this woman comes up to us, and she said, "Your child is very beautiful and has an unusual face," something like. She said, "I would like to paint her," and my mother said, "Oh no. I'm sorry, she doesn't do that." It's very funny because eventually, I ended up modeling for art classes. Most of my friends turned out to be painters, and many, many friends were jazz musicians, and the least were the poets.
Crawford:	I'd like to go back to something we touched on earlier. There are many allusions to black people in your poems. So that your poetry is like a collision of poetry with black culture –
weiss:	Yes.
Crawford:	In the forward to *Can't Stop the Beat*, I'm quoting: "Black people – dancers, painters, poets, and musicians – who have appeared throughout my life making deep impressions, here are some of those stories." (weiss 2011, 8) What is the black theme, the theme in your life, in your poetry? Where does that come from?
weiss:	My first connection with a black person was in fourth grade. Hitler has already started his destruction, but he hasn't reached Austria yet. He's still in Germany. I'm in fourth grade, and one day, a black girl appears as a new –she's part of our class. A teacher introduces her, and then we would work playing in the courtyard on a break,

| | and she was on one side of the court and I was on another and we immediately connected.

After school we'd start walking home together, and she just lived a few blocks after I did, but I noticed that she had a bruise always somewhere, her face, this, and I sort of found out enough that her mother was white, and very violent, and her father was South African, but he had split. |
|---|---|
| weiss: | We had a real nice comfortable time being together, and she was the first black. I'd never seen a black person. Okay. So this is 1937, and in the meantime, we've left Vienna, we went back, went to New York, different things happened. I'm in Chicago and I go to a party on the South Side, and I'm the only white person there, and this little guy who was very, very black said, "You're from Vienna, aren't you? I'm into semantics and I really hear you, and you have the same accent as a friend of mine from Vienna." I said, "Okay, pardon, is she black?" He said, "Yes," and I said, "She's tall and she has kind of light black skin, and she's a dancer?" He said, "Absolutely right. She lives just a block away in an attic." This was like 4 a.m., and we leave the party and go over there and there she is. We made all kinds of plans. I said, "How did you escape?" She said, "I joined an Italian circus and traveled with them." |
| Crawford: | That is quite a story! Well, let's move on to Chicago. Your parents returned to Germany after the war to work—your father as an accountant, and your mother worked for free to help the survivors as a translator. She asked you to help during a vacation. |
| weiss: | Yes. Everybody would say, "Do you know so and so in New York?" It was horrible what they were going through; I couldn't stand it. So anyway, so, when we got back in 1949, I finally took hold of my own life, left home, didn't know how to even earn a living. I found this building, this place called the Art Circle, near the North Side, and I rented a room there for seven dollars a week, twenty-eight a month, and painted my room black and all that. And, one day, I heard some jazz upstairs, and anyway, a friend of mine, this Ernest Alexander, who's black, came running down and said, "Come on; upstairs there's music." I said, "I'm too busy." I was in the middle of typing up a poem, and he grabbed the poem, pulled it out of the typewriter, and read it, then he said, "Hey," and I said, "Where are you going with –" He took the poem, went upstairs where they were jamming up there, and they were all black, and he said, "Hey guys, |

	you've got to hear this." So, they stopped playing for a minute, and then my friend Ernest started reading what I had written, and they just started playing behind it. You know, while my friend was reading, they started jamming instead of stopping and listening.
Crawford:	You were one of the first to fuse contemporary jazz with poetry, and you began that at the Art Circle in an informal way. Were your parents resistant to your going to the Art Circle?
weiss:	They had gotten me another room somewhere, and it wasn't much more, and they paid for it, but when I left that – it was a very boring situation – and I moved into the Art Circle, they stopped paying me!
Crawford:	They weren't okay with it.
weiss:	No.
Crawford:	You have mentioned here a long list of composers and music that made an impression on you. It's an interesting list to me. Do you remember what was on it?
weiss:	Yes, sure.
Crawford:	I have the list: "Bird, Prez, Lady Day, Bud, Monk, Villa Lobos, Django, Sarah, Schoenberg, Shostakovitch, Bartok." (weiss 2011, 14)
weiss:	Part of it was jazz, of course. I know Sarah Vaughan and Billie Holiday, those were the singers.
Crawford:	And Bud, Bud means Bud Powell?
weiss:	Bud Powell, right.
Crawford:	Bebop, modern jazz.
weiss:	Oh, bebop, the rhythms of bebop is so much like my – have you noticed, like my poems?
Crawford:	Oh yes.
weiss:	And I didn't pattern it after them. I just realized we had the same kind of rhythm, and improvisation was part of it. Now I also had other friends like Gerhard Samuel. Do you know who he is?
Crawford:	Yes, a composer and conductor who championed contemporary music during his time with the Oakland Symphony. His family came in the late 30s to escape the Nazis as well.
weiss:	Well, we were very close friends; very, very close friends, and I stayed with him several times in Seattle, and he also took one of my *Desert Journal* days and turned it into music and we had jazz, saxophone, and violin, and what do you call it when you have a triangle. Okay, and we had a rehearsal. That whole thing was supposed to be for the Monday Evening Concerts series at the museum, and yeah, he

took that and set it to music, all written out, so it wasn't improvisation.

Crawford: That Monday night series was quite special. But I'm interested in this list that I have in front of me, because you named Schoenberg; you chose Schoenberg's music.

weiss: Yes.

Crawford: You had Shostakovich, picked out Shostakovich, and Bartók.

weiss: Yes.

Crawford: What of Bartók did you especially hear?

weiss: I can't tell you. I have the records, I still have some of the records.

Crawford: That's impressive. Well, so, you are talking about Art Circle now and you worked at the Capitol Lounge, where you heard Percy Heath and Milt Jackson and the rest of the Modern Jazz Quartet.

weiss: It's just a little bar, famous. A few years later, maybe '50, maybe the second or third year of the Monterey Jazz Festival, and I was down there, and there was a large party afterwards, and Diz was there, and I went up to him and I said, "You probably don't remember me, but I was working at the Capitol Lounge as a dice girl." He said, "Oh yeah, I remember you."

Crawford: You told me you liked being a waitress and working in clubs because it was just like the theater! Well, we're focusing on the Art Circle now, and so I'd like to talk about some of the people were important in that particular era. You write about Ken.

weiss: Ken Holloway, yes.

Crawford: "KEN has a car. he is blue-black. we walk. talk. logic. / facts. mathematics. i love mathematics. // [...] i love to look at KEN. he's put together just so. and / his voice. low. held in check / [...] we meet often. it is important. then never again. / that also was important." (weiss 2011, 16) You say here, "i love mathematics"? I think they moved you from the Bronx school to the Harlem school because you knew algebra, isn't that right?

weiss: Well, we had quite a fling, and he was a mathematician and he was black, yes, and brilliant. I took four years of math, the last year I think was solid geometry, and I'm the only one in the class who got all A's for four years, and I was the only girl in the class.

Crawford: Do you think your mathematics affected your work?

weiss: Yes, both my rhythms which I aligned to the jazz, bebop, and mathematics, I think had a lot to do with music for me.

Crawford: Were you writing a great deal then?

weiss: Yes, mm-hmm.
Crawford: I read somewhere that you had written an autobiography that you didn't like; you destroyed it.
weiss: That was when I was in eighth grade, and it was really kind of stupid.
Crawford: I doubt that. You were a teenager already.
weiss: It was *Tomboy at Boarding School*. I remember going into detail about a volleyball game, about rules. Other people write about tennis, or football, or whatever, boxing, and I wrote about volleyball. [laughter]
Crawford: With the time left I want to talk about some of your work. A few weeks ago, the Pacific Film Archive in Berkeley showed *The Brink*, 1961, the first film you made. Steve Seid, who introduced the film, said it was "built around the existential musings of two contentious lovers," and "jettisons narrative logic for a skeptical embrace of the moment – the Beat moment." (2019) In the film you see a couple embracing on the beach, then in other places, a sweet nude scene with classical poses, and finally a vivid romp with your dog Zim Zum. What would you say of the narrative, if there is one, and how was the film put together?
weiss: The film was a collection of unexpected moments, an upside-down message on a bus, streets in Chinatown where I used to walk at 3 or 4 in the morning when no one was around. My artist friends did everything. The direction was, "This is what should happen. Take it from there." They were not acquainted with the poem; just told to show up in certain clothes.
Crawford: All of your music is improvised too. I've talked to Hal Davis, who accompanies you often, and Doug O'Connor, one of your bass players, and both say that they simply listen as you speak and play accordingly. But I'm wondering about a recording you made, *A New View of Matter*, with a young German cellist named Matthias von Hintzenstern; how that worked out.
weiss: I was invited to perform in Erfurt, Germany, on the way to Austria. When I got to the club, all the musicians had the flu. This was October, 1999. The person who arranged it was a member of a literary group. Matthias shows up and we set up and suddenly I realized he did not speak a word of English. I could communicate some things but he just listened to my voice. It was the first time I improvised with someone who spoke no English. He really picked up on what I was saying.

Crawford: Remarkable! In the music there are voice and instrumental overtones, a lot of harsh sounds with no particular pitch; it sounds very surreal to me. There's a second album that reminisces about North Beach called *All that Jazz,* and the musicians perform as you read. The notes indicate it was in 1993 at Kimball's East.

weiss: I was invited to perform for twenty minutes. I took excerpts from *Can't Stop the Beat.*

Crawford: You had no rehearsal.

weiss: No, just told the musicians to follow me. At one point the piano strings broke, and we played with the bass and drums.

Crawford: Larry Vuckovich on piano was a marvel in terms of musical mood, I thought. He does these dark riffs on the black, rough water of the Bay as you describe taking the ferry in your poem.

weiss: I agree. I spent a lot of time on the ferry. I'd go on the last ride over and back – Oakland to San Francisco. I went almost every night. The crew waved me on. The ferry ran until almost midnight, and there weren't many people.

Crawford: The image of you writing away on the Bay ferry in the darkness is a good one, and a good place to end today's interview. Thank you, ruth.

Works Cited

Seid, Steve. "On the Brink of Something: ruth weiss as Filmmaker." *UC Berkeley Art Museum and Pacific Film Archive (BAMPFA)*, 2019. Web. https://bampfa.org/page/out-of-the-vault-essay-ruth-weiss-brink. Accessed 25 January 2021.

weiss, ruth. *Single Out.* Mill Valley: D'Aurora Press, 1978.

weiss, ruth. *Can't Stop the Beat: The Life and Words of a Beat Poet.* Studio City, CA [Los Angeles]: Divine Arts, 2011.

weiss, ruth. *A Fool's Journey/Die Reise des Narren.* Bilingual English/German. Trans. Peter Ahorner and Eva Auterieth. Vienna: Edition Exil, 2012.

Frida Forsgren
ruth weiss and Visual Art: The Watercolor Haiku Series *A Fool's Journey* and *Banzai!*

Introduction

Compared to the relationship between poetry and jazz, that between poetry and visual art remains understudied in Beat culture. And as an individual genre, Beat art has received less attention from scholars than Beat literature and poetry.[1] One might say that the compartmentalization of Beat culture into separate genres is insufficient due to its multimodal and experimental nature: painters wrote, poets painted, and musicians accompanied words and images. Paintings were exhibited in venues where poems were performed to jazz. ruth weiss' body of work is indeed a characteristic example of a Beat oeuvre consisting of film, poetry, painting and music. Her decision to also paint her written, spoken and recorded haiku poems shows a willingness to experiment and to enhance the text's aesthetic possibilities. This chapter presents her two painted haiku series *A Fool's Journey* and *Banzai!* to explore the connections with the visual art scene in San Francisco in the postwar years. I propose that weiss' watercolor haikus display affinities with the style, themes and energy of Californian Beat art, as well as being striking examples of multimodal and improvisational art in American artistic counterculture. This chapter explores to what extent weiss' art may represent independent and individual artistic expression thus placing more emphasis on weiss' visual language, and less on her poetry and music.[2]

ruth weiss

Jewish Austrian poet ruth weiss settled in the US permanently in 1939 after having escaped the Nazi regime in Europe. weiss had begun to write poetry in Swit-

---

[1] Thomas Albright in *Art in the San Francisco Bay Area* (1985) refers to the visual artists connected to the Beat Generation as Funk artists. Recently, however, the term Beat Art is a more accepted term. See for instance Forsgren (2008) and Wickizer (2010).
[2] For a discussion of the visual poetry in ruth weiss' works, see chapter 4 in Estíbaliz Encarnación-Pinedo's dissertation *Beat & Beyond, Myth and Visual Arts in Women of the Beat Generation* (2016). Here the influence of visual art in her poetry, the visual aspect in her poetry and weiss' own involvement is presented.

zerland after the war during a one-year sojourn at a boarding school in Lausanne 1946/47, spending time hitchhiking and writing. Back in the US she settled in Chicago, where in 1949 she moved into the Art Circle – a housing community for artists where she began to experiment with poetry and jazz. This was also where she first recited her own poetry in front of an audience, and invented Jazz & Poetry. In 1950 she left Chicago and hitchhiked first to New York's Greenwich Village and the French Quarter in New Orleans, before going to North Beach, San Francisco, where she arrived in 1952 and continued jamming and reading poetry with street musicians.[3] weiss says she came to San Francisco before the whole Beat thing happened: "It was in full bloom I would say by 1955, but I was already in San Francisco in 52 living in North Beach. I was already living there when the other Beats arrived." (qtd. in Antonic 2014) It was during this time that she began to publish in Bob Kaufman's magazine *Beatitude*, and in Wallace Berman's *Semina*, and to contribute to the emerging Beat scene. In this same period, several of her musician friends (Sonny Nelson, Jack Minger, Wil Carlson) opened a club called The Cellar, in which she would regularly hold poetry and jazz sessions every Wednesday night. In addition, she made weekly appearances every Monday at the Coffee Gallery. She was the first poet to perform poetry to jazz, but other well-known male Beat poets such as Jack Kerouac who recorded their work have traditionally been recognized as inventors of the Jazz & Poetry scene. weiss left North Beach for Big Sur in 1957, and her pioneer role was underplayed in Beat scholarship for many years.[4]

Ever since the 1950s weiss has been one of the most prolific and active Beat poets, who continued to write and perform poetry to jazz all her life. In addition to her extensive catalogue of published poetry, and poetry and jazz performances, weiss' work includes film, theatrical plays, and watercolor haiku. ruth weiss is indeed a poet closely linked to visual art and to visual artists, as Encarnación-Pinedo notes (371). In *Can't Stop the Beat* (2011) she says: "most of those close to me are sculptors, painters. / my words carry pictures" (62). And her close relationship with artists such as Madeline Gleason, Aya Tarlow, photographer and poet Anne McKeever, dancer Fumi Spencer, painters Marianne Hahn, Mel Weitsman, Paul Beattie, Sutter Marin and Paul Blake indeed attest to this. In a 2002 interview she further elaborates on her own involvement with visual art: "I have done plays, I've done paintings, I've done stories, I've done theatre pieces, but I consider myself a poet whose other work – paintings, etc. – are simply an

---

[3] These biographical notes are based on the chapter "ruth weiss. The Survivor" in Brenda Knight (1996) and the interview with ruth weiss in Nancy M. Grace and Ronna C. Johnson (2004).
[4] See Thomas Antonic (2015), where he tells the story of how ruth weiss has used the "Beat brand" as a concept to relaunch her career in the Beat canon.

extension of the poem." (qtd. in Trigilio 2003) Her involvement in visual and interdisciplinary projects are several: California painter Sutter Marin illustrated weiss' book of poetry *Gallery of Women*, and Paul Blake illustrated *Desert Journal* (1977). Her narrative poem *The Brink* (1961) was turned into a film incorporating "found objects" and the improvisational aesthetics of Beat filmmakers. Paul Beattie was the cinematographer of the film. Moreover, ruth weiss has responded poetically to visual art. She has written ekphrastic poetry in response to art in the first section of *South Pacific* (1959), and in response to the work of sculptor Richard Yaski ("Rickard Yaski, his Sculpture 1991") (see Encarnación-Pinedo, 375). This multimodal approach has affinities with the work of several Beat poets and artists. The Beat milieu in California was dominated by a great will to experiment and to not let the creative process be limited to one single form or one single mode of expression. The visual work of Jack Kerouac, Lawrence Ferlinghetti, Allen Ginsberg, William Burroughs, and the *Semina* group poets and artists are well known, but the multimodal art of ruth weiss has received less attention. I suggest we first look at the postwar art scene in San Francisco to understand the visual culture that weiss' poetry develops in, and that might have had an impact for how she paints her haikus in the late 60s and onwards. The examples I present are from the 1980s and 1990s but may be said to embody and channel the energy of the San Francisco Beat Generation.

The San Francisco Art Scene

Art in California in the postwar period was infused with an intense feeling of energy and freedom, with spontaneity and ease, as if everything were possible and as if everything might be open for exploration. Artists were eagerly embracing abstract expressionism, while others were painting in a figurative style, yet others made collages, assemblages, light-sculptures, food art, political art, or organized quirky happenings and performances.[5] Kenneth Rexroth has compared the San Francisco art scene to Barcelona before the war due to the pulse and the energy of expression (Natsoulas, 41). It was in the air to look at everything anew and to re-invent art, its forms, its colors, its message, and its potential for expression. After the Second World War, the shared feeling was that one needed to start over again, and that one started with a blank canvas.[6]

---

[5] See for example the introduction in Thomas Williams (2013).
[6] "Postwar artists were encouraged to develop a free-form aesthetic: a modern art that abandoned the stylistic conventions of the past and was predominantly self-reflexive, focused on the qualities of its particular medium and the 'feelings' of its individual makers." (Doss, 125)

Moreover, the San Francisco art scene was clearly a global scene. The political and social situation in post-war Europe had led to an influx of artist immigrants entering US art milieus, of which indeed ruth weiss herself is an example of. Major expressionists such as Mark Rothko, Ad Reinhardt, Clyfford Still and the Picasso-pupil Jean Varda taught at the California School of Fine Art. The avant-garde expressionist Hans Hofmann had taught summer school classes at the University of California, Berkeley, in the early 1930s. Frida Kahlo and Diego Rivera stayed in the city in the early 1930s and the latter executed frescos at the California School of Fine Arts. The surrealist Marcel Duchamp lived in LA and the Dadaist Clay Spohn had shown his Dada-installations at the California School of Fine Arts in 1949. Moreover, California was open to the oriental influences of Asian cultures from across the Pacific mixing the quiet reflections of Zen culture with the energy and pulse of expressionism, seen for instance in the energetic pottery art of the time, as well as the interest in calligraphy and the haiku genre.[7] And at the newly founded San Francisco Museum of Modern Art students were exposed to the most advanced work of Cubism, Dadaism, Surrealism, and Expressionism. The student crowd in the art schools was dominated by war veterans who had used their Servicemen's Readjustment Act, the so-called GI Bill, to enter art school with free tuition – they were a tough crowd with serious life-experience from the battlefields in Poland or the Pacific Jungle, and they had a great urge for expression. This mix of sophisticated European avant-garde artists, ex-soldiers bursting with energy, and the intense hunger for new kinds of expression, turned the art milieu in San Francisco into a veritable powerhouse of creativity. This energy was furthermore clearly fueled by the Beat movement, which was an active agent in the city from the beginning of the 1950s with its underground galleries, poetry readings, performances and art happenings.

The traditional view is that whereas artists on the East Coast were preoccupied with the conceptual and intellectual conception of art, West Coast artists just *did it*.[8] And, indeed, several artists describe this lack of theoretical sensibility and intellectual awareness on the West Coast.[9] David Park states that "he needed something *out there* to serve as focus for his painterly energies" (qtd. in Boas, 135), which resulted in his break with abstract expressionism. And Joan Brown saw San Francisco as "an art city without a sense of past history

---

[7] Daniel Belgrad discusses the relationship between Pottery and Zen in California in *The Culture of Spontaneity: Improvisation and the Arts in Postwar America* (1998).
[8] See the chapter "San Francisco and the Abstract Expressionist Revolution" (Williams 2013) and "On the Farther Shore of Abstract Expressionism" (Landauer 1996).
[9] For descriptions of the informal character of the Beat milieu see, for instance, Albright (1985), Belgrad (1998), Freedman (2005), and Phillips (1995).

as New York." (Brown) In California, art was in a way "invented anew", and artists could choose their own personal paths. This non-dogmatic and experimental attitude on the West Coast favored freedom of expression and is also deeply reflected in how the art milieu was organized. Everything was allowed, and everything was interesting, it was not what you made or how you made it that was important, but that you *made* it. As Bruce Conner describes it, "The main thing was to make it, to make the image. To make the thing that you were trying to do, and whether it fell apart or not was of secondary importance." (Conner) What you made, how you made it and who you were, were factors of less importance. Arthur Monroe commented that

> Nobody cared if the painting lasted until tomorrow because NOW was the most important thing. Today and the experience of painting were the most important things. So, people poured everything they had into it. They wanted to discover what painting could mean for them with a tremendous feeling of urgency. (Monroe 1998)

The art milieu was organized spontaneously in the local underground art galleries,[10] and, as underscored by Joan Brown, people had an infectious energy: "It had to do with the energy at that time. We were all electric and in a hurry, in this terrible hurry. I don't know where we were all going. But we were in a big hurry to do something". (Brown) In this big hurry, capturing the moment was an essential part of an artist's style and expression. And not thinking, or talking about it, but doing it, getting it down on paper. There are obvious parallels between the free, spontaneous, improvised aesthetics of the postwar art milieu and weiss' poetics. Thomas Antonic argues that "spontaneity and improvisation have a major function" (2015, 193) in weiss' poetics. And he emphasizes that this was something she had already "developed long before getting in touch with any of the so-called Beat poets". Antonic furthermore argues that

> weiss' writing is more related to a free improvisational form of jazz, meaning that as the first step in the creation of a poem she focuses on a subject, and aims to write down words as uncensored by rational thought as possible, similar to the technique of *écriture automatique*. (Antonic 2015, 193)

It is also important to emphasize that Beat poetry and abstract expressionism were only two expressions of a free, spontaneous, improvised aesthetics that

---

10 See the catalogue *The Beat Generation Galleries and Beyond* (Howard 2006) for a discussion of the underground art scene in the Bay Area. Here the Dilexi, the Spatsa, the Six, the East-West, the Batman and the King Ubu are presented thoroughly.

characterized the postwar avant-garde. In *The Culture of Spontaneity* Daniel Belgrad has shown how such diverse impulses as bebop jazz, gestalt therapy, Black Mountain College, Jungian psychology, experimental dance, and Zen Buddhism had one unifying theme: spontaneous improvisation. And in addition to being a spontaneous, improvised culture, another characteristic feature of postwar aesthetics is that it very often consists of multimodal projects. Very often poets paint, painters write, or dancers, painters and musicians collaborate on aesthetic projects. And just as often individual artists embark on projects that involve poetry, painting, sculpture and music. Let us look at three examples from the Californian postwar art milieu to see how this tendency made artists embrace a range of styles and expressions. In this milieu it is difficult to talk about one coherent artistic style – we are rather dealing with complex multimodal expressions clearly driven by principles of spontaneity and improvisation.

A case in point is visual artist George Herms who was active in the crowd around Wallace Berman's Semina Gallery in the late 1950s and early 1960s. Herms first wanted to become a writer but became one of California's leading funk and assemblage artists. The oil painting *Larkspur Does a Hemorrhage*, the woodcarving *Nalota*, and the poem *Circumstances Around Larkspur Does a Hemorrhage*, all made after the birth of his daughter Nalota in 1960 is an example of a multimodal creative process. The birth of Nalota took place in the couple's Larkspur home with Herms as midwife, and inspired three different works of art. The painting *Larkspur Does a Hemorrhage* is a vertical canvas with two main shapes: a dynamic red flame in the lower part of the composition and a stable circular shape in the upper part. The energy of the flame and the stability of the circle fuse two diverse cosmic energies, and the physical presence of blood from the birth make it a strong, personal expression. The painting has a figurative pendant in the expressionistic woodcarving of Herms' wife giving birth, and the poem *Circumstances Around Larkspur Does a Hemorrhage*, with its cyclic structure and strong poetical imagery, evokes the circumstances around the birth. The powerful impact of birth made Herms both write a poem, paint an abstract artwork and make a wood-carving, and shows the same kind of willingness to let an idea or a feeling find its realization in different forms.

A different case in point is Fred Martin, a painter from the milieu around the California School of Fine Arts, who had a playful, improvisational and spontaneous way of working. Martin exhibited landscape sketches at the Six Gallery in October 1955 during Allen Ginsberg's famous reading of "Howl", sketches that had been made as he drove through San Francisco at day and nighttime. Martin says that he drove around the city looking for sights to paint, mounting little pieces of Masonite on the steering wheel of his car, stopping to paint when he found the inspiration. The paintings are like a diary of the artist's forays throughout the city, and the technique of sketching has affinities with the kind of sketching encouraged by Jack Kerouac in his "Essentials of Spontaneous Prose".

A final case in point is from the oeuvre of Bay Area's most acclaimed visual artist Jay DeFeo and her monumental *The Rose*. *The Rose* is a striking example of an artwork that evolved through different stages, and through different media. The massive painting made on DeFeo's living room wall between 1958–1965 can be interpreted both as a painting, as a sculpture, or as a performance piece. Furthermore, it plays the lead protagonist in Bruce Conner's film *The White Rose* (1965), which tells the story of when *The Rose* was removed from the artist's studio. In the assemblage piece *The Stool*, paint from *The Rose* is incorporated in the sculpture, and one-year DeFeo broke off pieces of the painting to be used as small elements in Christmas cards to her friends.

These three examples all testify to the multimodal, experimental and spontaneous visual culture in the Bay Area that I suggest ruth weiss' production has strong affinities with. The way that her poetry work "pours over" into music and into painting has a similar experimental approach. However, in the passages quoted before ruth weiss says that her words carry pictures and that her paintings are simply extensions of the poems. But is it possible to say something about her visual art as art? Let us look at her painted watercolor haikus as examples.

ruth weiss' Watercolor Haikus

ruth weiss began writing haiku poems in 1953 or 1954 while living at the Wentley Hotel in San Francisco. During this time, she encountered Jack Kerouac, and the two hung out and began to write haiku together. They were both interested in setting haiku to and against jazz music and formed a method of dialoguing in haiku between themselves, writing haiku back and forth to one another.[11] Jack Kerouac was a great admirer of ruth weiss' haiku and said that they were superior to his, but none of these early texts survived. A haiku is a traditional Japanese poem that does not rhyme, traced as far back as the 9th century. It consists of three lines, with the first and last lines having five "moras," and the middle line having seven (referred to as the 5-7-5 structure). A mora is a sound unit, much like a syllable, but is not identical to it. Since the moras do not translate well into English, the haiku has been adapted to where syllables are used as

---

11 None of these haikus survived. See, however, the track "Writing Haiku with Jack Kerouac" on weiss' Audio CD *Jazz & Haiku* (weiss 2018).

moras. In Japanese, haikus are traditionally printed in a single vertical line, while haiku in English often appear in three lines to parallel the three spoken phrases of Japanese haiku. ruth weiss' haikus do not follow any strict pattern, but are written both as two, three or four lines in the visual representations. A haiku is more than a type of poem; it is a way of looking at the physical world and seeing something deeper, like the very nature of existence. It should leave the reader with a strong feeling or impression. ruth weiss has said that "the haiku is a fabulous discipline for making each word succinct, meant, cutting out the fat, a perfect exercise for poetry. By cutting out the fat, making it bone" (Grace and Johnson, 65). In the late 60s ruth weiss started painting her haikus in watercolor. And since 1965, she has exhibited her water-colors on at least three occasions: In 1965 at Joker's Flux Gallery in North Beach, where she did a group show; in 1980 in a solo show entitled *Banzai!* at Gallery Become (San Francisco), and in 1994 in a show called *A Fool's Journey* at Mendocino Moulding Gallery (see Encarnación-Pinedo, 398). In what follows, I consider her two painted series *A Fool's Journey* and *Banzai!*. My analysis is divided into a description, followed by a more detailed analysis of aesthetic elements/affordances as the water-color technique, color contrasts, and figure style, and how the haikus relate to zen culture.

A Fool's Journey: Description

A Fool's Journey consists of nine haikus forming a narrative sequence. The poems were written in the sixties, painted in 1994, published in 2012 (text only), recorded in 2017, and (re-)published in 2018 (for the first time as paintings).[12] The poems tell the story of the physical and existential travels of a meandering soul with the characteristic short, rhythmic haiku language:

> one foot off the cliff / the other on solid rock / her heart in her mouth
> down your beat wild wind / rocking to your tune all night / fool smiles into sleep
> in search for her word / she travels book in hand / the pages are blank
> the dance in the flame / the picture in the embers / tells all ever known
> there—there the bridge is / every so often
> is this really dark / no moon the milky way
> MA EARTH stop quaking / the fools are encircling you / with love returning
> layer by layer / she peels the onion she is / and laughs with her tears

---

12 See weiss' *A Fool's Journey/Die Reise des Narren* (2012) for the poems and *Flugschrift* (2018) for the recording and reproduction of paintings. Six of the paintings in the series belong to ruth weiss, while three were sold and are in a private collection in California.

Night breaks into dawn / a fool sees stars in the day / red-twinkling in the sun
(printed in a different sequence in weiss 2012, 14)

When ruth weiss paints the poem, she makes use of a similarly poignant, "to the point" aesthetics as she does when she writes. In the first haiku painting, which corresponds to the first three lines in the poem, she paints a small figure which has one foot on a cliff and the other in mid-air. A close-up of a female face in the sky gives the setting a contemplative mood (see Img. 1). The second haiku is the only abstract painting in the series. Here the wild wind is rendered by undulating grey lines on a blue background. In the third painting the traveler is gone, but we see a little book with blank pages blowing in the wind over a vast turquoise sea. The white words are painted on the brown cliffs. In the fourth painting she paints a close-up of a fire encircled by white stones, the yellow words are written on a deep blue background (see Img. 2). The fifth haiku shows a bridge in the sky that casts reflections in the sea. The white words are written on a blue background. The sixth haiku shows a hand stretching towards the stars and the milky way. The hand, the milky way, the stars and the words are white on a dark blue background. Only one patch is bright red. The seventh shows mother earth as a round orange cat with ears, squinting eyes and a line of people holding hands around the middle. The words are black on an orange and blue background. The eighth haiku shows a red onion and a red onion peel painted on a white and blue background. The red text is written on a light blue background. The ninth haiku shows a red and blue bird and a yellow sun on a red-twinkling sky. The text is red, except for the word *night* which is blue, the paper is kept white. As in the poems, we note how the visual vocabulary is deliberately kept simple and "to-the-point" with no excess features.

Fig. 1: Haiku painting from the series *A Fool's Journey* (1994)

Fig. 2: Haiku painting from the series *A Fool's Journey* (1994)

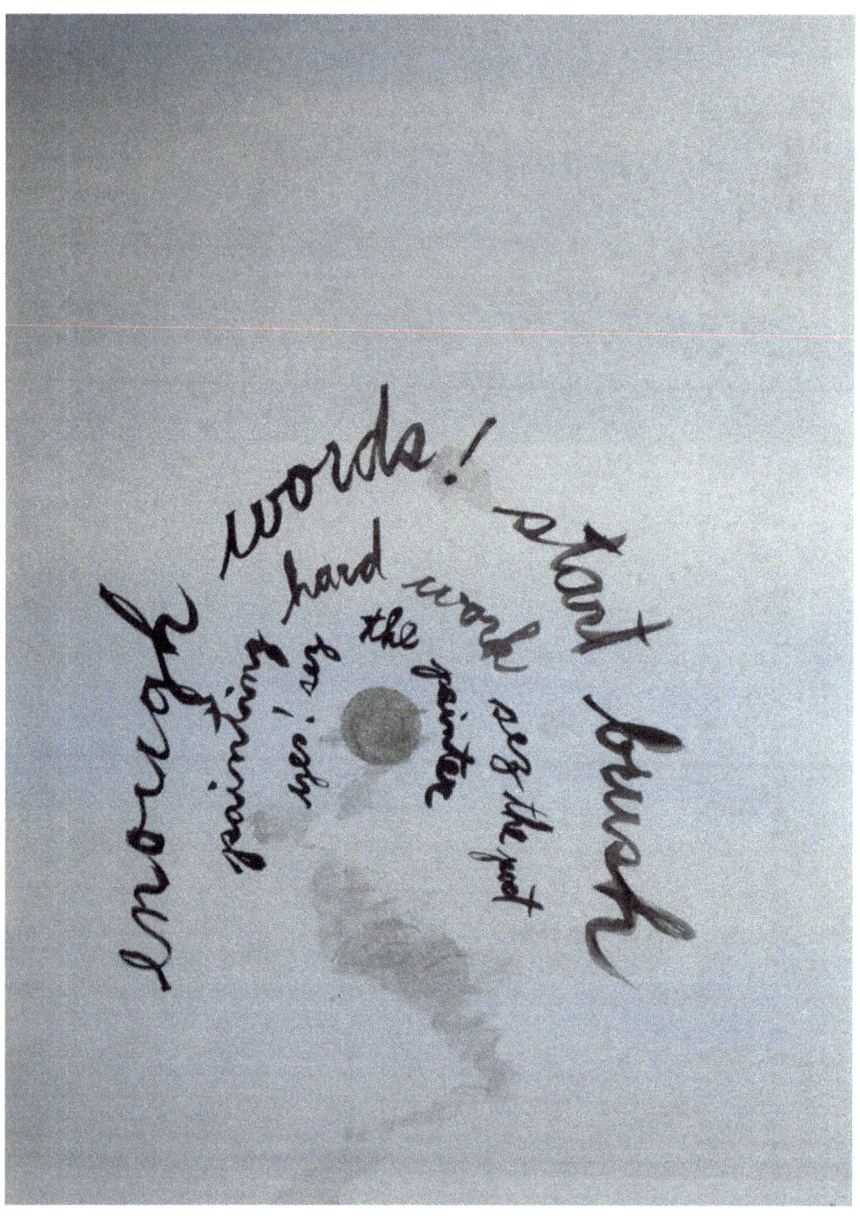

Fig. 3: Haiku painting from the series *Banzai!* (1980)

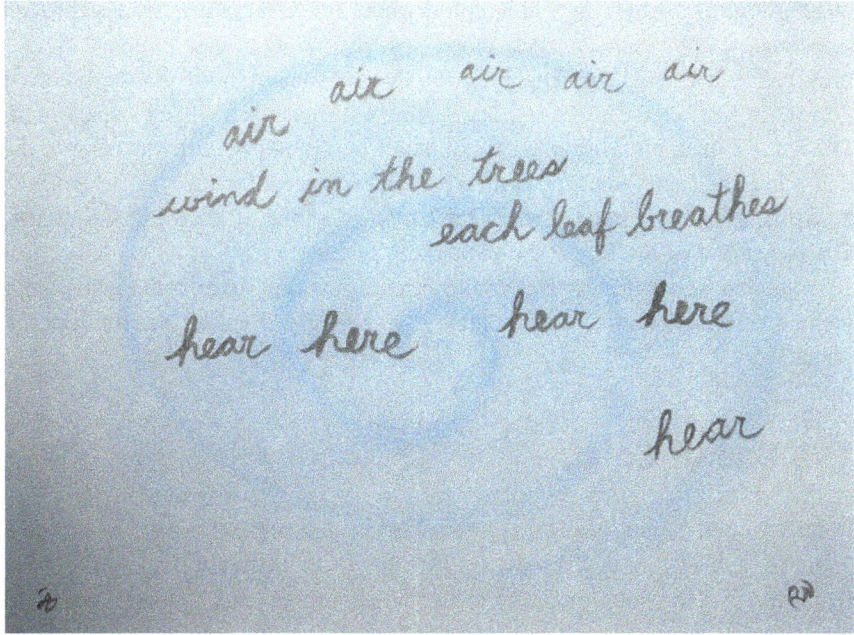

Fig. 4: Haiku painting from the series *Banzai!* (1980)

The *Banzai!* series: description

The *Banzai!* series consists of 25 haiku poems written and painted in the early 1980s.[13] From the series I have chosen to analyze the five following haikus:

> enough words! start brush / painting hard work sez the poet / yes! sez the painter
> air air air air air / wind in the trees / each leaf breathes / hear here hear here hear
> a mask of each mood / each nuance that brushes us / and each is the real
> red dog of my past / still in current memory / your name is ZIMZUM
> RULE OF THE HOUSE – NO PETS! / so the white cat from next door / is the live-in guest
> (the first three are printed in weiss 2018[14])

The "enough words" haiku consists of dark blue letters on a white background. The words are distributed in a circular pattern, diminishing towards the center where a circle closes the composition (see Img. 3). The "air air air" haiku consists

---

13 See weiss (2018) for the recording and reproduction of some of the paintings.
14 The latter two haikus are unpublished and archived in the ruth weiss Papers.

of black words written on a blue spiral where the white paper shines through (see Img. 4). "A mask for each mood" consists of two dark points and four blue concentric lines opening in a fan-like movement from the center. "red dog of my past" consists of a red dog casting red shadows on a grey background. The name Zimzum is written in capital letters and in red echoing the color of the dog, while the rest of the text is grey. The first sentence in "RULE OF THE HOUSE" is written as a sign, while the white cat is painted with only a few black contours. The text and sign are black; the paper is left white.

ruth weiss uses aesthetical compositional features such as the water-color technique, color contrasts, and figure style to expand and enrich the written poem.

The Use of Color

weiss' painted haikus are characterized by a delicate lightness. The specific aqueous, soft feeling of the watercolor technique creates a transparent quality that enhances the simple quality of her poetic lines. Paint is never applied in dense layers, so that it will allow the light and porous quality of the paper to shine through. She uses few colors carefully tuned to harmonize, not to contrast with each other. In the series *A Fool's Journey* the dominant colors are different varieties of blue, ranging from a dark blue in "is this really dark" to a light white-blue nuance in "one foot off the cliff". In the latter she uses a darker blue tonality in the writing, and in the former she uses white to allow text and background to gently complement each other. Only in a few painted haikus does she use strong colors to create more intense effects. In "is this really dark" a sharp red star emerges from the blue background, and in "the dance is the flame" the strong yellow bonfire and the yellow text create a sharp contrast to the blue background. A gentler tone and inverted contrast are expressed in the haiku "in search for her words", where the white writing seems a mere whisper on the sepia sandy cliffs. The strongest color contrasts are used in the "MA EARTH" haiku where mother earth is painted in a strong orange, and in the "layer by layer" haiku where the onions are dark red and heart shaped.

In the examples from the *Banzai!* series we also see a similar soft use of color. This series, too, is dominated by a harmonious use of blue, grey, and white backgrounds, with the colors of the writing tuned to gently contrast with the background. In some examples, she introduces more contrastive elements, as for example in the "red dog of my past" haiku, where the strong orange color highlights the dog as a magical presence emerging from the grey and misty background. Or in the "enough words" haiku where the darker contrast created between black words and sky-blue background makes the words seem even more forceful. And in "NO PETS!" the delicate black outline makes the contours of the white cat emerge strikingly. But an overall compositional feature in these haikus is that text and images complement each other, making the written words blend in with the painted background.

The Use of the Figures

ruth weiss paints in a naïvely simple figurative style. Things from the physical world are carefully rendered concretely as they appear in the poems, and she does not attempt abstraction. The tiny balancing figure in "one foot off the cliff" is painted as a human figure with one foot off the cliff and the other on solid rock, the tiny book blowing in the wind in "in search for her words" is rendered as a book with blank pages physically blowing in the wind, and the bonfire in "the dance in the flame" shows a bright bonfire encircled by rocks. In "is this really dark" the hand reaching into space stretches towards a white, dotted milky way, and the bridge in "there—there" has the shape of a concave bridge casting reflections in the sea.

In essence, the haiku is a concise and concrete form of poetry; ruth weiss underlines how this form allows her to "cut out the fat, making it bone". Only the words that need to be there are written down to create a concise, condensed composition. And when she paints her haikus, we see that she uses a similar type of concise, concrete aesthetics. When she writes about the flame, she paints a concrete flame, when she writes about the milky way, she paints a concrete milky way. When she paints, she shows us "the thing" (the dog, the bridge, the book, the flame) in a similar way to what she does when she writes about the thing or the phenomenon. Her paintings are governed by the same kind of "cut out the fat, making it bone" aesthetic: the figures are simple and to the point, the colors of figures and words are in tune, and the watercolor technique ensures a quality of lightness, ease and spontaneity. As in the written haiku, a further symbolic or deeper meaning in the picture is expressed by how the aesthetical qualities such as color, color-contrasts, texture, figures and composition are combined. The tiny figure on the verge of the abyss in "one foot off the cliff", the lonely hand stretching out into the vast universe in "is this really dark" and the vibrant flame in "the dance in the flame" all function as strong images that carry a deeper existential meaning. In addition, a characteristic feature in ruth weiss' haikus is how the simplicity of the words is underlined by the simplicity of the images. And the combination of the low-key word/image aesthetics makes her paintings a subtle harmonious entity that allows a reflection of the deeper meaning of the text.

The Influence of Zen Stylistic Simplicity

This way of stripping the picture down and letting it consist of simple concrete elements has stylistic affinities not only with the simplicity of the haiku form, but also with Japanese Zen aesthetics. In meditative Zen painting the artist makes the calligraphic sign or the mandala in an intensely focused movement or breath. We may say that in Japanese Zen painting the sign *is* the picture. And if we look at the style of several of weiss' haikus they have a similar kind of concentrated energy. In some of the haikus from the *Banzai!* series the visual sign approaches an even more symbolic shape than in the figurative and narrative compositions of *A Fool's Journey*. In the haiku "air air air" the inward-moving blue spiral focuses energy towards the center and the distribution of the words accentuates the spiral shape. In the haiku "enough words" the text is distributed in a concentric movement around a central shape, echoing a mandala. The words are larger in the outer circle, diminishing as we approach the center. Also, the circular water encircling the Zimzum dog carries connotations with the mandala, encircling him and framing him as the central figure. And in "a Mask for each mood" the mask consists of two simple dots placed in a concentric fan pattern, again echoing the circular mandala.

As in meditative Zen calligraphy, ruth weiss also makes use of a style in which the most significant feature in the composition is both centered and highlighted. She thus draws inspiration from an art form that captures the moment with concentrated energy and focus. The presence of calligraphy, Buddhism and Zen teaching had been an important influence in the Bay Area ever since the early 50s with the teachings of D. T. Suzuki and Shunryu Suzuki. "San Francisco in the '50s was a Mediterranean and Asian city, in temperament and culture" (qtd. in Kowinski) as the poet Michael McClure recalls. McClure further elaborates: "We encountered Asian people every day, signs in Chinese, markets in Chinese, food in Chinese and Chinese music in the street." (Kowinski) The influential Black Mountain College theoretician and musician John Cage had attended D. T. Suzuki's lectures in 1945, scholar and writer Alan Watts broadcasted on Zen at KPFA Berkeley from 1953, and many of the Beat poets such as Gary Snyder, Philip Whalen, Allen Ginsberg, Jack Kerouac, Joanne Kyger, Diane di Prima and Michael McClure all became dedicated practitioners (Falk, 254). In the visual Beat milieu, too, the fascination with Zen art was strong as we see for instance in the Buddhist and Zen inspired work of Beat artists Michael

Bowen, Arthur Monroe and Keith Sanzenbach. Sanzenbach made monumental mandalas on burlap, and ephemeral mandala ink drawings on rice paper. Arthur Monroe experimented with action painted calligraphy, and Michael Bowen painted Zen monks and mandalas.[15] ruth weiss' interest in the mandala shape, her delicate watercolor haikus and written body of poetry attest to how much she absorbs the presence of Zen and incorporates it in her work. Moreover, ruth weiss' first husband Mel Weitsman, who was a painter and had studied under Clyfford Still, later became a Zen priest and was with Shunryu Suzuki the co-founder of the Berkeley Zen Center and its abbot until 2020.[16] Weitsman's presence in her life further intensified the influence of Zen on her art.

Conclusion: ruth weiss' multimodal and improvisational aesthetics

Encarnación-Pinedo states that "[t]he haiku's concise language and juxtaposition of images or ideas not only falls comfortably within weiss' poetics, but also within her interest in the visual arts (397). She argues that weiss' visual art enables her to "show" rather than to "explain" (373). But weiss' involvement in visual art is more than an interest. This chapter shows that weiss' watercolor haikus are placed firmly within Beat culture's multimodal aesthetics. By experimenting with watercolor weiss manages to give the poems, in addition to their verbal and musical form, also a concrete visual form. And by adding the visual mode, she extends and expands the original text in a similar way to that in which George Herms and Jay DeFeo extended their ideas beyond one style or one format. weiss' haikus are written, painted and performed, thus showing an expansive and playful creativity.

This chapter furthermore argues that ruth weiss' watercolor haikus fit well within the culture of spontaneity in postwar California by being improvisational and sketch-like. Her spontaneous, improvised style in the watercolor haikus has affinities with the kind of "just doing it" aesthetics that characterizes the West Coast art milieu. The example of Fred Martin who drove around San Francisco while using the steering wheel as an easel is a parallel example of the shared improvised aesthetics that characterizes postwar art on the West Coast. Fred Martin sketched the city as he drove by, while weiss sketched the images in her haikus, trying to capture the specificity of the concise, concrete Japanese po-

---

15 For information about the artists and illustrations see Forsgren (2008).
16 Hakuryu Sojun Mel Weitsman, born July 20, 1929, passed away on January 7, 2021.

etics. Moreover, the watercolor technique provides the paintings with a transparent feel that underscores a light, improvised quality. Lastly, the concise way she paints "the thing" from her haikus mirrors the precise Japanese aesthetics she emulates. The connection between the written word and the painted image, has similarities with how the painted sign in Zen painting *is* the image. The painted haiku becomes a concrete way of showing or revealing the message to the beholder, capturing its essence in a visual form.

Works cited

Albright, Thomas. *Art in the San Francisco Bay Area: An Illustrated History*. Berkeley: University of California Press, 1985.

Antonic, Thomas. "Vienna Never Left My Heart: A Conversation with ruth weiss, Recorded and Transcribed by Thomas Antonic." *European Beat Studies Network*, 2014. Online: https://ebsn.eu/scholarship/interviews/ruth-weiss-interviewed-by-thomas-antonic/ Accessed 25 Jan. 2021.

Antonic, Thomas. "From the Margin of the Margin to the 'Goddess of the Beat Generation': ruth weiss in the Beat Field, or: 'It's Called Marketing, Baby.'" *Out of the Shadows: Beat Women Are not Beaten Women*. Edited by Frida Forsgren and Michael J. Prince. Kristiansand: Portal, 2015, pp. 179–199.

Belgrad, Daniel. *The Culture of Spontaneity: Improvisation and the Arts in Postwar America*. Chicago: University of Chicago Press, 1998.

Boas, Nancy. *David Park. A Painter's Life*. Berkeley: University of California Press, 2012.

Brown, Joan. Oral history interview with Joan Brown, by Paul Karlstrom, July 1 – September 9, 1975. Transcript on 35 mm microfilm reel no. 3196. Archives of American Art, Smithsonian Institution, Washington DC. Unpublished.

Conner, Bruce. Oral history interview with Bruce Conner, by Paul Cummings, 16 April 1973. Transcript. Archives of American Art, Smithsonian Institution, Washington DC. Web: https://www.aaa.si.edu/collections/interviews/oral-history-interview-bruce-conner-12017#transcript. Accessed 20 Jan. 2021.

Doss, Erika. *Twentieth-Century American Art*. Oxford: Oxford UP, 2002.

Encarnación-Pinedo, Estíbaliz. *Beat & Beyond: Myth and Visual Arts in Women of the Beat Generation*. University of Murcia, PhD dissertation, 2016. Web: https://www.tdx.cat/bitstream/handle/10803/369842/TEEP.pdf?sequence=1&isAllowed=y Accessed 25 Jan. 2021.

Falk, Jane Elizabeth. *The Beat Avant-Garde, the 1950s, and the Popularizing of Zen Buddhism in the United States*. Ohio State University, PhD dissertation, 2002. Web: https://etd.ohiolink.edu/apexprod/rws_etd/send_file/send?accession=osu1486461246817201&disposition=attachment. Accessed 25 Jan. 2021.

Forsgren, Frida. *San Francisco Beat Art in Norway*. Oslo: Forlaget Press, 2008.

Freedman, Stephen. *Semina Culture: Wallace Berman and His Circle*. Santa Monica: Santa Monica Museum of Art, 2005.

Grace, Nancy M., and Ronna C. Johnson. *Breaking the Rule of Cool: Interviewing and Reading Women Beat Writers*. Jackson: UP of Mississippi, 2004.

Howard, Seymour, et al. *The Beat Generation Galleries and Beyond*. Davis, Ca.: John Natsoulas, 1996.

Knight, Brenda. *Women of the Beat Generation: The Writers, Artists, and Muses at the Heart of a Generation*. Berkeley: Conari, 1996.

Kowinski, William S. "Buddha by the Bay / Eastern Religion Meets the West Coast Art Scene." SFGate. Web: https://www.sfgate.com/bayarea/article/BUDDHA-BY-THE-BAY-EASTERN-RELIGION-MEETS-THE-2832169.php. Accessed 25 Jan. 2021.

Landauer, Susan. *The San Francisco School of Abstract Expressionism*. Berkeley: University of California Press, 1996.

Monroe, Arthur. "The Decade of Bebop, Beatniks and Painting." *Somarts Cultural Center*, 1998. Web: http://www.somarts.org/beat/beat_text2.html Accessed 24 Oct. 2018–10–24 [now offline, saved version: https://web.archive.org/web/20180226061312/http://www.somarts.org:80/beat/beat_text2.html {Accessed 25 Jan. 2021}].

Natsoulas, John. *Lyrical Vision: The 6 Gallery 1954–1957*. Davis: Natsoulas Novelzo Gallery Press, 1990.

Phillips, Lisa. *Beat Culture and the New America: 1950–1965*. New York: Whitney Museum of American Art, 1995.

Trigilio, Tony. "An introduction." *The E-Poets Network*, 2003. Web: http://voices.e-poets.net/weissr/intro.shtml. Accessed 20 Jan. 2021–01–20.

weiss, ruth. *A Fool's Journey: Poems and Stories / Die Reise des Narren: Gedichte und Erzählungen*. Bilingual English / German edition. German transl. Peter Ahorner and Eva Auterieth. Vienna: Edition Exil, 2012.

weiss, ruth. *Flugschrift*. Special issue (no. 25) of *Flugschrift*, edited by Thomas Antonic and Dieter Sperl. Vienna: Literaturhaus Wien, 2018. Print + Audio CD [entitled *Jazz and Haiku*].

Williams, Thomas. *The Bay Area School: Californian Artists from the 1940s, 1950s and 1960s*, London: Lund Humphries, 2013.

Image Sources

Img 1–4: Haiku paintings by ruth weiss, first published in *Flugschrift*. Special issue (no. 25) of *Flugschrift*, edited by Thomas Antonic and Dieter Sperl. Vienna: Literaturhaus Wien, 2018. Print + Audio CD [entitled *Jazz and Haiku*].

Lars Movin
"Go to the roundhouse, he can't corner you there": *The Brink* (1961), ruth weiss' Poetic Film

> we are of man and woman
> and having named us
> both will still
> use
> us as battlefield
>
> we scream
> to find our balance
> (ruth weiss, "Blue in Green", 1960)

In December 1952, when the French filmmaker and author Jean Renoir introduced his 1939 movie *The Rules of the Game* (*La Règle du jeu*) at a New York screening organized by Cinema 16, Amos Vogel's now legendary avant-garde film club, he stated that this "is a normal picture about abnormal people". So, how about the subject of this essay, ruth weiss' poetic vision from 1961, *The Brink*, a 40-minute black-and-white exploration of the state of things between the sexes? Not exactly a "normal picture". And definitely not about "abnormal people". Then, what is it? *An abnormal picture about normal people?* Well, perhaps. Or maybe rather something completely beyond categorization: an original take on humanity, as seen through a unique blend of poetry and images, sound and music, voice and movement.

When interviewed on the subject, weiss usually insisted on *not* being a filmmaker, but rather a poet who sometimes used other media or tools in her ongoing internal journey toward self-discovery. She might paint her words on canvas, she might work with music (as in Jazz & Poetry), she might try her hand in theater or performance, she might act in works by other people (as has been the case in a number of films by Steven Arnold), and she might make her own films (the one discussed here, *The Brink*, as well as one later collaboration, the montage *Las Cuevas de Albion*, co-directed by weiss and Alfonso Gordillo in 2001); but no matter what form of artistic expression she has been engaged in, it was always from the perspective of a poet.

This manner in which *The Brink* came together unscores this fact. In 1960, shortly after having completed a long narrative poem entitled "The Brink" (published in 1978 in the collection *Single Out*), ruth weiss was approached by a friend, painter Paul Beattie, who had bought a 16 mm movie camera. It was

https://doi.org/10.1515/9783110694550-025

just about the time when the new phenomenon of *underground film* was emerging, with people like Maya Deren, Marie Menken, Kenneth Anger, Stan Brakhage, Ken Jacobs and others coming together in something that felt and looked like a movement.

Throughout the 1950s, jazz and poetry (separately or together) had been the two primary means of expression in the neon-lit urban nights, the bonfires around which subcultures would gather. But with the arrival of the 1960s it became time to visualize the dreams, longings, feelings and ideas otherwise suppressed by mainstream society. Jazz had helped liberate the body, as poetry had the mind and language, but now it had become time to render taboos visually. And what medium could be more perfectly suited to storm the citadels of conventions and control than *film* – the very reflection (*and* sometimes source) of the American dream?

By 1960, the photographer Robert Frank and the painter Alfred Leslie, both based in New York, had just finished their collaborative movie *Pull My Daisy*, a loose interpretation of a play by Jack Kerouac and with his somewhat improvised narration on the soundtrack, a film that not only turned out to be *the Beat movie per se*, but also became the sparkle that ignited what is known as the New American Cinema Group. And on the West Coast, more precisely in San Francisco where weiss had her home at this point, the local communities of poets, artists, musicians and the like had borne equally original and innovative films such as Christopher Maclaine's *The End* (1953), Vernon Zimmerman's *Lemon Hearts* (1960) and Ron Rice's *The Flower Thief* (1960), the latter a showcase for future Warhol Superstar Taylor Mead and with weiss in a small part which, however, ended up on the editing-room floor.

In short, Paul Beattie couldn't have timed his acquisition of a movie camera better. With the arrival of economically accessible lightweight 16 mm cameras, celluloid dreams were in the air, and weiss didn't have to think twice about accepting her friend's invitation to write a script. Before going to work, however, she mentioned to Beattie that she actually already had something that might lend itself to the big screen, namely the aforementioned poem, "The Brink". Why not simply make a film of that?

Part 1

In "The Brink" – the poem – we meet a woman and a man, she a poet (perhaps) and he a painter, a couple tentatively circling each other in what seems to be a night-long rendezvous at a cafeteria, but could also be a series of encounters, or even the anatomy of a relationship over a longer period of time. The text is fragmentary, disjointed, jump-cutting from one image, sentiment, situation to the next, moving in circles with no clear indications of space and time.

What we get is a feeling of fluidity and uncertainty, hope and disillusionment. Everything in the poem's universe seems to be latent and immanent rather than tangible and in full bloom. And quite fittingly the general metaphor is water. The scenes are set against the ocean with its ebb and flow; the water moves constantly moving yet always stays the same. It is a space with a blank surface and mysterious depths, and with seemingly endless stretches of water that nonetheless end at the shore.

As for the characters, they are equally enigmatic, both realistic and archetypical. The "She" and the "He" could be a specific couple – like ruth weiss and her then-husband, painter Mel Weitsman – but they could also be any woman and any man, trying to reach each other under the weight of the eternal night of humanity. The cafeteria is clearly a concrete location in the poem, but at the same time the text repeatedly hints at in-between places: between day and night, land and sea, past and future, reality and dream or imagination: "like anyplace in the world of heaven" (weiss 1978, n.p.). Thus, the mode of the poem is transitional, the state of being is brimming with possibilities, and everything is seen as potential. As the film's title suggests, the characters seem to exist in a world which is on the brink of becoming a place that allows for a real and equal meeting between the sexes, and for the existence of love in the deepest and truest sense of the word. But it is only on the brink. It's like the metamorphosis will never reach its ultimate destination. The tadpole will remain a tadpole, the caterpillar a caterpillar. Or will they? The poem ends: "butterfly / she said / beyond" (weiss 1978, n.p.). Perhaps there is hope after all.

In the film we get representations of all of the above, but of course in a different form. The couple is personified as a real "She" and a real "He", actors certainly, but still flesh and blood. The cafeteria materializes in front of our eyes, anonymous, but yet real. The same goes for the other locations and characters – real places and people that could be seen in the San Francisco of 1960, but at the same time universal symbols of the condition of mankind. Yes, it's all there, right down to the theme of metamorphosis – in the film firmly established in the very opening shot of a caterpillar happily nibbling away on a leaf.

In the film, we also get the sensation of being witnesses to a situation that's on the brink of becoming something else. A collapsing world order leaving space for a new and better world? A looming disaster that might pave the way for a new sensibility, a new understanding, a new language, a new balance between the sexes? Or just some undefinable way of existing that is on the verge of being born, not once and for all, not again, but born continually into the present?

In order to reduce *The Brink* – the movie – to being a mere visualization of the poem, however, would be unfair. The film *is* that, but it is also so much more. Of course, thematically speaking, the poem and the film deal with quite similar subjects. And the poet/filmmaker herself has been kind enough to provide a few thoughts on the matter, naturally in the form of a poem (printed on the cover of the DVD version of the film):

> the caterpillar is a phase
> in the development of the butterfly.
> it can crawl to the brink
> but it cannot yet fly.
>
> the human being in his complexity
> moves toward completion
> through the many doubled selves
> all stranger & familiar
> moves through toward completion as lovers
> and the he & she fearing love as completion
> call forth all interference
>
> this until the simple point of contact
> where the butterfly is possible
> (weiss 1961, n.p.)

Again, we have the optimistic vision in the end – the possibility of the butterfly – but before reaching that point the two lovers have to navigate their way through a darkness of fear and uncertainty. And this is not only their own individual respective darkness as woman and man, it is a much larger darkness that concerns the state of things on a grander scale, the general condition of humanity and the post-war madness of the world, which in one way or another served as a common ground for most of the Beat Generation writers and their peers.

So, what is it we encounter in *The Brink*? A love story? Kind of, but not really. A statement on the war between the sexes (to paraphrase Leonard Cohen)? Not quite. An exploration of the conditions for communication or maybe even love between the sexes? We're getting warmer. A statement about the necessity of constantly being prepared to take up the challenge of trying to become human in an inhumane society? Ever warmer. Which brings us back to the beginning – the notion of watching an unusual movie about two people who, given all their imperfections, appear to be quite 'normal'.

Let's take a look at the making of *The Brink*. The process went more or less as follows: Scanning the poem for visual motifs and possible scenes, ruth weiss began scribbling down ideas for locations and situations in a notebook that would then serve as a storyboard or logbook for the shooting of the film. Then a team of filmmakers and actors was assembled, mostly friends and acquaintances. With weiss herself as the natural choice for the director, Paul Beattie, the owner of the camera, would serve as cinematographer, while the composer Bill Spencer would be responsible for the soundtrack. This was it, more or less, in the technical department, a level of modesty that is typical of the no-budget ethos of most underground movies, reflecting both the limited resources and the artistic ideal of being able to move freely, be spontaneous, improvise, act and *re*act. *The Brink* was a cinematographic poem, but it was also like a saxophone solo blown weekend after weekend all over the Bay Area until all of the film's sixteen, seventeen, eighteen (depending on how you count) scenes were in the can.

As for the cast, in the two main parts, as the nameless couple, the "She" and the "He", weiss picked two painters, Lori Lawyer and Sutter Marin. And in casting the extras, more friends and acquaintances were drawn in, including the filmmaker herself, her husband, and even their dog, ZimZum; or maybe one shouldn't say *even*, as the dog also appears in the poem – "and zz barked & barked" (weiss 1978, n.p.). When the time came for the shooting – which usually took place on Sundays, as some of the involved had regular jobs – Beattie would pick up weiss in his VW bus, and together they would make a round to pick up the other actors and crew members and go to the location of the day. In finding the suitable locations, weiss would draw on her observations as an obsessive nocturnal wanderer. Since arriving in San Francisco, weiss said, she made a habit of going out for long walks after dark, just drifting around, exploring various neighborhoods, getting the feel of the city, and looking for places to sit down and scribble in her notebook, either on a park bench or in a bar or coffee shop. In this sense *The Brink* is both a documentation of San Francisco at a certain time in history, and an autobiographical statement, a sort of mental map, an exteriorized reflection of the psycho-geographical layout of the poet's mind.

Among the chosen locations, a handful spring to mind: The cafeteria in the opening scene, which serves as a stand-in for the Foster's on the ground floor of the Wentley Hotel on Polk and Sutter Street, a Beat hangout where weiss herself had lived in the mid-'50s – the actual cafeteria in the film was located near the Civic Center on Market Street and Van Ness Avenue and was, according to weiss, chosen because of its many mirrors. The streets of Chinatown, a popular place to roam for many artists and writers, Beat and otherwise. Sutro Heights Park near the Cliff House on San Francisco's Pacific coast – selected for its statues, especially the now-gone "Satyr's Dream". The bunkhouse on the waterfront of the former codfishery in Belvedere-Tiburon, rented by artist couple Dave Lemon and Jerry O'Day, just north of San Francisco, with a spectacular view of the Bay (see Seid 2019). And a newly constructed neighborhood of tract houses in Pacifica, a coastal community a few miles south of San Francisco.

On principle, weiss stated, none of the actors would know what would happen in their scenes on any given day. They would be told what to wear or what to bring, but the scenes they would appear in had to be spontaneous and improvised, based on some guidelines given by the director – a mood, some kind of interaction, either intimately between the woman and the man or within the context of a social situation, such as a public space or with a party of friends. After the shooting was completed weiss did the editing herself. And with the visuals of the film in place, she went back to the narration – read by herself – and cut and adjusted it to fit with the images, a process that also involved inserting fragments from other poems or texts – and even weaving parts of different poems into each other. "The Brink" remained the backbone of the narration, but in between segments of this long poem are heard parts of the poems "Blue in Green" and "Chopsticks" (both 1960), "Light" (from the 1976 collection *Light and Other Poems*), "Earth Painting" and "Words No Words for a Dance Fumi" (both from the 1959 collection *Gallery of Women*, the latter being the only poem to appear in its entirety), and even some lines from a play by weiss called *Figs* (first produced in 1965).

Finally, there was the soundtrack, a clever construction by Bill Spencer with additional music by Warner Jepson and Mel Weitsman, not so much a selection of songs or pieces of music as a *total composition*, a soundscape running for the length of the film, shifting between a jazz improvisation, (an illusion of) incidental music – such as flute playing – and more abstract elements, including acoustic effects that seem to be environmental sounds but are in fact created to function as a composition à la *musique concrète* – an example being what appears to be the sound of the ocean but is in fact, according to weiss, made by rolling one or more metal chains over a floor (weiss 2017).

A further note on the music: As a pioneer of Jazz & Poetry, weiss is naturally associated with jazz, but one of the stylistic aspects that makes *The Brink* stand apart from many other American avant-garde films of the time is that the use of actual jazz on the soundtrack is limited to only one segment. The movie itself can in certain ways be said to be permeated by a *bebop poetics* – the use of improvisation, the mood and feeling of the scenes, the movements and interactions of the characters, etc. – but speaking strictly about the score the jazz element is certainly overshadowed by subtle and moody pieces of an avant-garde or improvised nature, at times with an oriental flavor.

All of the above combined lends the film a feeling of being both authentic and constructed simultaneously, like a vérité observation of real people in real places, but at the same time a documentation of real people *performing as them-*

Fig. 1: Sutter Marin and Lori Lawyer in *The Brink*

selves in real places. Again, like a sax player improvising on a theme, following the chord changes of a specific tune but also expressing him- or herself and thus creating something entirely new. As if to say: This is life, but it is also a language, a representation, a poem, an artistic expression.

Part 2

There is quite a clear structure to *The Brink*, not in a linear storytelling sense, however the film does follow a poetic, dream-like logic, moving from one situation to the next, from one stage of a relationship to the next, from one mental and/or emotional state to the next. The narrator's voice is that of a woman, not necessarily the female protagonist, but nonetheless a voice talking from a – primarily – female perspective, verbalizing observations, emotions and expectations of a woman entering a relationship. But it's also a reflection on the battle of the sexes from the point of view of an outsider, or a neutral observer, even at times entering the mind of the male protagonist. An example of the latter being when the narrator at one point, after "He" has arrived by sailboat to a wooden house by the sea, comments: "go to sea, she can't corner you there" – a mirror-

Fig. 2: Lori Lawyer in *The Brink*

ing of one of the film's key lines (appearing both in the opening scene and towards the end): "go to the roundhouse, he can't corner you there". As if both "She" and "He" were struggling with the dilemma, the conundrum of love: how to maintain your own individual freedom while at the same time giving it up for the Other? It's not so easy: "to live in paradise / is more difficult than not" (weiss 1978, np).

In the beginning of the film the two protagonists meet at a café where they sip coffee and converse and flirt with each other, surrounded by reflections, both in the form of actual mirrors and in the form of other people acting similarly – among them an elderly couple that may be seen as a premonition of the two protagonists' future, whether that might be considered a promise or a warning – according to weiss the elderly couple was one of the many gifts of chance they experienced during the shooting of the film (weiss 2017). The scene raises the question of what we are looking for in a relationship: Are we searching for the Other, or just a confirmation of ourselves? And it serves as an entry into the – potential – love story, a starting point for the journey of getting to know each other, exploring the field of emotions and attractions growing between the two.

"Go to the roundhouse, he can't corner you there": *The Brink* (1961) —— **187**

Fig. 3: Lori Lawyer in *The Brink*

At this stage, the rest of the world seems to disappear into the background, and the couple now exist in a virginal limbo between now and eternity, between innocence and desire, between what was and what might come, as they play and circle each other, first on a beach – "by the sea, by the sea" – then by a small river in the woods, then again on a different beach. The couple is increasingly absorbed by its two-ness, absent-minded like children and both only with eyes for the other, but yet – as the narrator stated in the opening scene – in a more existentialist sense they and their peers are also "all in this alone together".

From the carefree Garden of Eden-like situation of the first stages of love, the couple gradually work their way back to society, now appearing in various social contexts in their newfound dual identities as both individuals and parts of a double entity. Symbolically the film at this point presents them as being reborn, both emerging from the ocean; "He" arriving by sailboat to a house by the sea, perhaps a future nest for their life together, and "She" stepping out of the water like a dripping Venus, although fully clothed.

Together they take the house into their possession, a walking bass, soon followed by other jazzy instruments, underscoring the improvised dance of their nascent relationship. And soon they find themselves surrounded by friends in what seems to be something in between a bohemian bacchanal and a modern

Fig. 4: Sutter Marin in *The Brink*

version of the Last Supper, but also could appear to be some sort of theatrical performance, a group of people all engaged in artistic expression, playing music – flute and zither – acting, dancing, dressing up, perhaps in a process of reinventing themselves in the context of the dawning of a new social reality, a new society, a new culture. The film then cuts to the woman and man exploring the streets of Chinatown, a location chosen perhaps to emphasize how 'exotic' and fresh everything now appears to them with their heightened sensuality and invigorated energies. The scene has certain absurd or surrealist connotations – "She" carrying a bird cage and a stuffed bird – but it also serves as a visual equivalent of a passage in "The Brink" (the poem) where weiss elegantly links the traditional role of women as caregivers with a liberated notion of female sexual appetite, nicely wrapped in Chinese food terms: "the won ton woman / carried the hot dish across town / under her cloak / to keep it hot / instead of flying or sent for / her neighbor she was sick // wanton woman / where were you at the supper? / i want my man" (weiss 1978, n.p.).

The following sequences, first in a bus – with the director, Hitchcock-like, in a cameo as an anonymous passenger – then at a street-fair with a merry-go-round and whirling dancing into the ecstasies of night, serve as a sort of interim in the relationship, maybe a pause for reflection, but also a slide into a darker mode while the lovers seem to lose their way, with the narrator's poetry now balancing Heaven and Hell, hope and despair, which is underscored by ominous music. Which again leads into an even darker scene, with "He" lying alone on a couch in a cheaply furnished room, watching vague shapes of the city outside through dirty windows, apparently in a depressed mode. In his daydreams, we are led to believe, the couple is back in the paradise of the early stages of its relationship, now represented by a park – a more cultured, civilized version of nature – but with an element of wounded male sexuality in the form of a battered, decapitated statue of "Satyr's Dream". A sign of a more demonic element having entered their relationship? Well, perhaps, but certainly a hint that they are now way past their initial innocence.

Where to go from here? They cross a bridge, "looking for apples with no fences / but the tree was gone // instead there were rows & rows / of boxes" (weiss 1978, n.p.) – the "boxes" being newly built tract houses, the very manifestation of the conformity of the family-oriented and materialistic American lifestyle. Having caught a glimpse of this potential future, so different from the individuality of the picturesque bohemian house by the sea, the couple returns to the location from the opening scene – "they returned to the cafeteria / and the mirrors / and the clanging / dishes that shouted / it's your jungle / keep it clean" (weiss 1978, n.p.). But somehow after this retreat they again find themselves by the tract houses, as if this is the only path forward society will allow them. They give it a chance and step into the image of a happy young couple, walking their dog along the rows of identical houses, an environment as barren as the surface of the moon. And of course, it doesn't work. Soon they find themselves having turned their backs to the mainstream vision of the society – the inhuman suburbia – and instead they start over, building their own little City Upon a Hill, or not quite a city, but rather a foundation for a new life, created out of the rubble of the old world. This is where the film leaves the "She" and "He", joyfully playing in the ruins of what was, caught in a post-apocalyptic chaos – "past autumn past fall / past falling" (weiss 1978, n.p.) – but nonetheless seemingly optimistic on behalf of themselves, love, and humanity.

In the final images, the woman and man are studying some collages, a new artistic expression put together of fragments of a discarded past, a new vision, a new language. They hold a white envelope in their hands, perhaps a letter to the future. And on the envelope is crawling the black caterpillar from the film's opening shot – a small furry creature, not exactly pretty, but hopefully on the brink of becoming a butterfly.

Part 3

One can only speculate where the impulses and aesthetics of *The Brink* came from. As mentioned above, the film premiered in the wake of early underground works such as Robert Frank and Alfred Leslie's *Pull My Daisy* (1959) and Ron Rice's *The Flower Thief* (1960), just as its timing could hardly have been better in regard to the rising of the New American Cinema phenomenon. But although *The Brink* was certainly thematically and stylistically related to other artistic films made by her contemporaries, ruth weiss does not seem to have been particularly inspired by American avant-garde film in general. When asked about this, she claims not to have seen works like *Pull My Daisy* before embarking on her own cinematographic adventure, though later had reservations about the film due to its "boys' club" ethos. And likewise, she doesn't remember having ever attended Frank Stauffacher and Richard Foster's "Art in Cinema" series at the San Francisco Museum of Art, a hot spot for Bay Area avant-garde filmmakers in the 1950s (see weiss 2017).

Generally speaking, there weren't that many places to watch underground films on the West Coast at that time. And if weiss went to the movies, which she eagerly did, she would most likely attend screenings of the masters of European or Asian art cinema, like Jean Renoir, Jean Cocteau, François Truffaut, Federico Fellini, Ingmar Bergman and Akira Kurosawa. These films, however, were not direct influences on *The Brink* either, weiss says, or if they were, only to a very limited degree. Rather *The Brink* grew out of weiss' practice and sensibility as a poet – combined with observations from the various artistic and literary communities in San Francisco that she met during the 1950s. Never actively seeking to be part of any group or movement, weiss met and got acquainted with many of the key figures of the San Francisco Renaissance and the surrounding artistic milieus.

She was friendly with poet Madeline Gleason, founder of the San Francisco Poetry Guild and director of the First Festival of Modern Poetry in 1947 and thus a central character on the Bay Area poetry scene. She knew poet Robert Duncan and painter Jess Collins and maybe once or twice visited the couple's house on 1724 Baker Street – in 1952–54 also home to filmmaker Stan Brakhage. She was certainly aware, she says, of Kenneth Rexroth's weekly salons on 187 Eighth Avenue – and later 250 Scott Street – one of the incubators of the San Francisco Renaissance. She regularly went to the artist-run galleries on Fillmore Street, especially the East & West Gallery, where her husband Mel Weitsman first showed his work, and the Six Gallery, where she met Paul Beattie, without whom *The Brink* would never have happened. She knew most of the people at "Painterland," an artist commune on 2322 Fillmore Street inhabited by, among others, Michael McClure, Jay DeFeo, Wally Hedrick, Joan Brown and Bruce Conner, the latter both an artist and a filmmaker. And most famously she was a fixture at The Cellar, a jazz club where in 1956 she was the first to read her poetry with jazz accompaniment, thus being one of the inventors of what would soon be labeled Jazz & Poetry (see Antonic 2015).

All of this combined with weiss' exposure to Beat figures such as Jack Kerouac, Neal Cassady, Bob Kaufman and Philip Lamantia, including everything they stood for, directly or indirectly was to be woven into the fabric of *The Brink*. But one could also just say that *The Brink* was simply born out of its time and place. More than anything the film is a snapshot or a documentation of the optimistic and yet fear-filled cultural climate of the Bay Area around 1960, a time marked by a sense of darkness and need for change. Just as the Summer of Love in 1967 was followed by more disillusioned and violent cultural currents, the high point of the San Francisco Renaissance – the mid and late 1950s – was followed by more negative aftereffects, including police brutality, various forms of suppression and a general decline in health and spirits by many of the forerunners of the cultural revolution.

A crack had opened in the American landscape, allowing a glimpse of a different future, more humane and diversified, providing breathing space for all the various strands in society, including otherwise marginalized groups. And weiss caught that moment in *The Brink*, but she also caught the breaking point of the wave and how it started to roll back, returning more or less to the status quo upheld by the establishment and political powers. Which is one of the main strengths of the film: the sense of being exactly in tune with the Zeitgeist. In watching *The Brink* one can get this sense of the film summing up what had been going on at that point in history, but at the same time pointing to various phenomena, including a shift in the mental weather, just before all of this was articulated more broadly. One example that could be mentioned is the scene with the tract houses near the end of the film, in which the narrator is talking about "rows & rows / of boxes" (weiss 1978, n.p.). A phrase, and an observation of the development of the suburban landscape of the 1960s, which brings to mind Malvina Reynolds's song "Little Boxes" with its catchy references to mass-produced houses made out of "ticky-tacky," and which "all look just the same". An obvious inspiration for *The Brink*, one would think, had it not been for the fact that Reynolds' song was written a year *after* the completion of the film and only made famous by Pete Seeger in 1963. Still, the connection is striking: the lyrics of the Reynolds song were inspired by the Westlake District, a section of the San Francisco suburb Daly City, just a few miles north of Pacifica, where weiss shot the tract-house scene for her film.

Conclusion

The Brink was first screened at the San Francisco International Film Festival in 1961 and soon got a certain amount of local repute, not least thanks to the efforts of Canyon Cinema, a newly founded filmmakers' cooperative run by Chick Strand and Bruce Baillie. But in spite of its premiere being perfectly in sync with the signs of the time, the film was widely ignored by the movers and shakers of the underground film movement on the East Coast. The first person of some nationwide stature in the American film community to really acknowledge the qualities of *The Brink* was Stan Brakhage, who in December 1962 praised it in a letter to Jonas Mekas, co-founder of the New American Cinema Group in New York and editor of the magazine *Film Culture*. The letter – which was printed in the Summer 1963 issue of *Film Culture* (No. 29) – was meant to be a report on the state of things on the West Coast but zoomed right in on ruth weiss and *The Brink*. Brakhage had attended a screening of the film the night before and in his opinion it "certainly should be included in any statement on S.F. individual films and filmmakers". Brakhage then went on to introduce some of the people involved in the making of the film, especially Paul Beattie and Bill Spencer, responsible for, respectively, the camera work and the soundtrack, and to conclude his praise he said:

> I am once more aware of the complete lack of attention to film endeavor in this community (that *The Brink*, one of the most ambitious 'first' films I've ever seen, attempting to pitch the actors into situations preordained by Ruth Weiss' poetry yet leave them free of the context, unaware of the poetic narrative intended, to develop synthesis of poetry and image highly structured but containing a residue of very real immediate, almost haiku, feeling – that this film remains, almost two years after its completion, almost completely unknown); and my determination to correct such local ignorance by way of a little national advertisement, if possible. (Brakhage 1963, 80)

The key words here are "leave them free", "synthesis" and "highly structured". Brakhage certainly had an eye for the way in which weiss in *The Brink* was working with all sorts of methods or principles that had to do with breaking free of conventional art formats, using improvisation, chance, serendipity, collage, found objects, etc., but all within a structure – like jazz, like certain forms of spontaneous poetry, and also like Abstract Expressionist painting. As for the second key word, *synthesis*, Brakhage was acknowledging the organic interplay of the various elements of the film – images (acting), voice (poem), music (sounds) – and how they come together and add up to something that is greater than simply the sum of the parts. Which is really what causes the film to be much more complex than the poem "The Brink": the way in which image and sound are constantly in a dialogue, feeding – and feeding off of – each other, sometimes in a supporting or amplifying way, sometimes juxtaposing or counterpointing each other, but always structured in a well-thought-out and interesting way.

Nonetheless, it took a few decades for the broader film, art, and poetry communities to embrace *The Brink* in the same way as Brakhage had done it in 1962. In fact, the film pretty much stayed in the shadows until 1995, when it was included in the first major museum survey of the Beat Generation, *Beat Culture and the New America, 1950–1965*, at the Whitney Museum of American Art in New York. In 1996 *The Brink* was shown at the film festival in Venice, Italy, and from that point forward, word started to spread, just as ruth weiss was also experiencing a renewed interest in her poetry, and in her role in a more general sense as a pioneering figure, not only of the San Francisco Renaissance, but as a – female – voice of the American post-war landscape.

In that broader perspective *The Brink* still stands strong as a statement on the importance of the enigmatic powers and magic of poetry and spontaneity, and the importance of constantly articulating alternatives to the dominant discourses of a suppressing culture, constantly questioning gender stereotypes, constantly seeking out new ways of expression, constantly challenging the status quo, constantly staying in motion not to risk being cornered in one of the many roundhouses of a square society.

Works Cited

Antonic, Thomas. "From the Margin of the Margin to the 'Goddess of the Beat Generation': ruth weiss in the Beat Field, or: 'It's Called Marketing, Baby'." *Out of the Shadows. Beat Women are not Beaten Women*. Edited by Frida Forsgren and Michael J. Prince, Kristiansand: Portal, 2015. pp. 179–199.
Brakhage, Stan. "Letter to Jonas Mekas." *Film Culture*, vol. 9, no. 29, 1963, p. 80.
Seid, Steve. "The Forgotten Artists of the Belvedere Codfishery." *Open Space*, 20 June, 2019. Web: https://openspace.sfmoma.org/2019/06/the-forgotten-artists-of-the-belvedere-codfishery/ Accessed 23 Dec. 2020.
weiss, ruth. Interview by Lars Movin, Albion CA, 2019. Unpublished.
weiss, ruth. *Blue in Green*. San Francisco, Adler Press, 1960.
weiss, ruth, dir. *The Brink*. 16 mm, 41 min, 1961. [Digitized version on DVD ca. 2000, self-produced and distributed by the poet; restored version by the Pacific Film Archive, Berkeley, 2019] Film.
weiss, ruth. "The Brink." *Single Out*. Mill Valley: D'Aurora Press, 1978.

Image Sources

Img. 1–4: Production stills from *The Brink*, digitized by Thomas Antonic in 2017. Courtesy of ruth weiss.

Steve Seid
ruth weiss, Luminosity Procured

Sometimes nudity is the perfect solution – any self-respecting bohemian can testify to the truth of that. The resources needed are what one carries, the shape of one's earthly form. And so it was that ruth weiss found herself posing for a mid-sixties life drawing class at the San Francisco Art Institute (SFAI). For a poet in struggling mode, these simple gigs paid for that café latte in North Beach's Coffee Gallery or maybe a month's rent for a flat in S.F.'s inner Mission area.

As happenstance would have it, the sometimes-nude model was descending a flight of stairs, having just exited a friend's studio in the Fillmore District, when on the ascent was a young SFAI art student, Steven Arnold, about to visit same. It's the classic "I didn't recognize you with your clothes on" encounter, but thoroughly lacking in embarrassment. The two, a late-thirties poet and an early-twenties artist, became fast friends and co-conspirators in sundry forms of artistic demolition. Steven was about to leave behind the painter's easel for a more mobile medium, experimental cinema, and was collecting like-minded people for the staging of his first filmed short, an allegorical romp to be titled *The Liberation of the Mannique Mechanique* (1967).

Born in Oakland, California, Steven Arnold was of true artistic temperament, not just an occupational hazard but an unshakeable inclination. As a mere child, Steven entertained his small circle of friends with puppet shows, weaving exotic and liberatory fantasies. Entering the San Francisco Art Institute, he was already strongly defined by a self-generated style, all-embracing beliefs, and eclectic lustings. He drew from Tantric philosophies, Rococo art, and a fluidity of gender identification that was pioneering. Interesting that this sui generis kid from the East Bay would find a sympathetic soul in ruth weiss, a green-haired Austrian émigré and Beatific poet.

ruth would star in several of Steven's student films, including *Various Incantations of a Tibetan Seamstress* (1968) and *Messages, Messages* (1969), though "star" belies the fact that these films rely on a cast subsumed by the director's all-encompassing vision, a vision in which actors have an allegorical weight that crushes individuality. As noted experimental film artist (and Steven's professor) Robert Nelson once exclaimed, "Even the bodies of the actors look as though they were designed for the film," as quoted in the Film-Makers' Cooperative catalogue. Nevertheless, ruth persisted with grace to add her unabashed presence to what critic Kate Wadkins called, "Arnold's [...] dream-like visions of androgynous beings. Their narratives are modern-day fairy tales and reveries about gender – all through the lens of an acid trip." (Wadkins 2011)

https://doi.org/10.1515/9783110694550-026

Fig. 1: ruth weiss in the materialization chamber

Fig. 2: ruth weiss in the materialization chamber

Not wanting to screen his art school thesis film, *Messages, Messages*, in a traditional academic setting, Steven rented a floundering movie theater on the edge of the City's Chinatown. Shown at midnight, along with classic European avant garde shorts, the screening drew multitudes, enough to get requests for a repeat performance. Soon the on-going midnight series of rarely-seen avant garde films, cartoons, and experimental shorts became known as the Nocturnal Dream Shows. As an added spectacle, a troupe of cross-dressing gender saboteurs known as The Cockettes became a kind of vaudevillian opening act. Within a few months, the outrageous, norm-busting Cockettes became the headliners, bumping the films from the marquee.

This was anything but discouraging for Steven Arnold because The Cockettes would become the core cast for his next and most ambitious film, *Luminous Procuress* (1971). Filmed almost exclusively in a former industrial laundry in San Francisco's Mission district, Arnold's hallucinatory frolic follows two young men who enter a subterranean zone in search of a priestess, the "Luminous Procuress," who will lead them on a journey of discovery. The Procuress is played by Steven's lifelong friend, the exotically comported Pandora with her "bird-like visage."

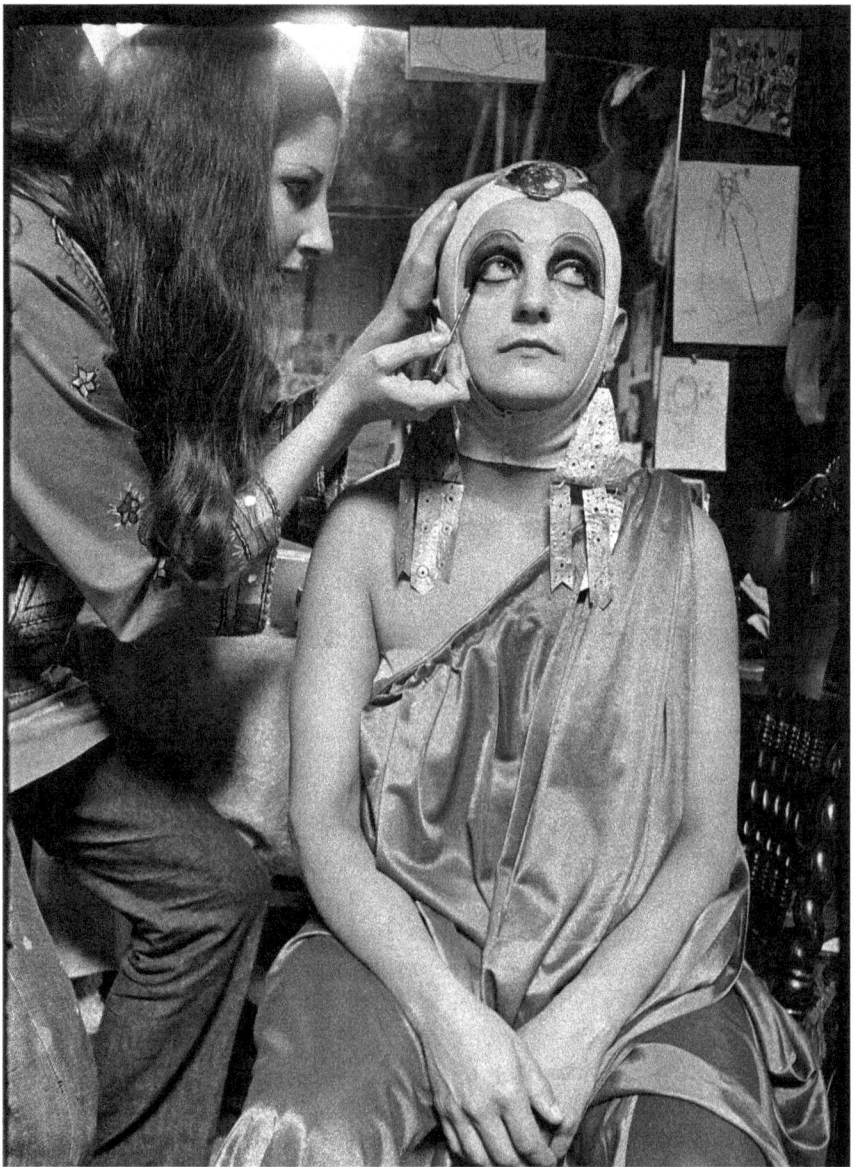

Fig. 3: Make-up artist Koelle with ruth weiss

The odyssey unfolds as an extravagantly designed series of "visionary tableaus" in which the young men slowly jettison their received notions of things and begin experiencing new possibilities for communion, sexuality, and finally

Fig. 4: ruth weiss in Golden Gate Park scene

Fig. 5: ruth weiss in the Quilt Room (by Scott Runyon)

being itself. Arnold's most prophetic contribution lies here, in his vision that identity is permeable, perhaps provisional, and that relinquishing the dualities of he/she frees you for a more enlightened existence.

To accomplish the ambitions of *Luminous Procuress*, Steven enlisted his faithful coterie of actors, artists, and anarchists, ruth weiss among them. The profusely-imagined look of the film called for equally extravagant costumes, make-up, and hair. And ruth was certainly game. Her hours before the make-up mirror and being fitted with Scott Runyon's inspired costumery were a greater commitment of time, then the filming itself.

Inside the allegorical narrative, ruth plays three separate characters, all unnamed. In one instance, she can be found, diaphanously robed, among what Steven calls "the witches, Pandora, ruth, and Lila King," three powerful enchantresses who are "pure alchemy, turning the air into clouds of jewels." They are the spiritual force that instigates change as guides through the "underground labyrinth" in the Procuress's realm. Later, in a rare outdoor scene, ruth will surface as the "second maid," wearing a loose-fitting purple dress akin to the garb of a temple attendant. There is frolicking in a shady glade while nearby a Sufi holy man is meditating. Eventually, "the witches" retreat to another part of this wood to form a magic circle, dancing gleefully. ruth, herself, romping in strange abandon.

Fig. 6: The three witches, Lila King, Pandora, ruth weiss (by Scott Runyon)

In another episode, ruth appears as a time traveler, clad in a glistening frock and elaborate mirrored diadem. She ritualistically moves a reflective orb up and down as she seems to materialize inside a light-refracting chamber. This moment connects directly to the film's closing scene in which the same chamber is occupied by what appears to be a transcendent otherworldly being.

When ruth next appears, it is in a "lush Chinese/Quilt Room," a sacred feminine sanctum within a larger contested space. Surrounded by magical "fairy godmothers," ruth, wearing a wiry headdress like an object of unusual power, reads from an undisclosed poetic source in an incantatory manner. The godmothers are rapt, and their mute attention is in opposition to the anarchic energy exuded by the ribald and reliably raucous scenes dominated by The Cockettes.

At this point in *Luminous Procuress*, the young lads are nearing their ontological ascension – they will soon be rewarded with magical powers. But nudging them along, the three witches, Pandora, ruth, and Lila, turbaned and scantily draped, will provide a sort of sirenic chorus, a cacophony of supra lingual soundings, that will heighten their transformation.

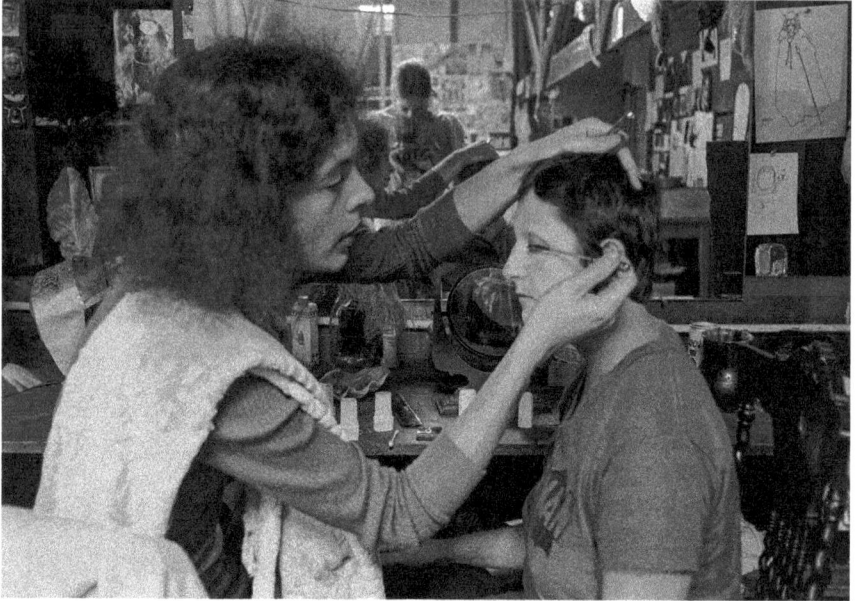

Fig. 7: Steven Arnold applying make-up to ruth weiss

Luminous Procuress would be the last film Steven Arnold would complete. And ruth would only collaborate with him one more time – in a play about the Hindu Monkey God, created by Steven and clothing designer Kaisik Wong. It was a charmed time for all involved, a time of unbridled experimentation, deep self-discovery, and a melding of emancipatory aesthetics.

Writing at the time, Steven Arnold pondered ruth his "poet and tragedian." He saw her as "pure moonlight. A magic talisman—living, positive, charmed ... She perceives and reflects my work perhaps more completely than anyone else. Her eyes tell you she knows all. Part Edith Piaf, part Giulietta Masina, truly one of the most important women on the planet. I wouldn't make a film without her."

ruth was the Luminous Procuress or, at least, a charmed prototype.

Fig. 8: ruth weiss awaiting make-up and costume

Works Cited / Image Sources / Notes

All photographs by Ingeborg Gerdes (courtesy of Owen Gump), except where noted. Photographs by Scott Runyon, courtesy of ruth weiss. Unreferenced quotes are from Steven Arnold's journal written during the making of *Luminous Procuress* and from a handwritten synopsis of which I seem to have the original.

Wadkins, Kate. "Projecting Female Identities." *Hyperallergic*, 7 July 2011. Web. https://hyperallergic.com/28209/projecting-female-identities/ Accessed 1 Dec. 2020.

In 2017, *Luminous Procuress* was restored by the Berkeley Art Museum & Pacific Film Archive in partnership with the Walker Art Center. Thanks to Ingeborg Gerdes and Scott Runyon for their photographs; Jon Shibata, the PFA Collection; Vishnu Dass of the Steven Arnold Archive; and, of course, ruth weiss.

Thomas Antonic
The ruth weiss Papers

The following chapter is intended to be a survey which provides an initial overview and information on the scope of ruth weiss' papers and should be of use for many aspects of future research on the poet. In 1992, weiss began selling parts of her papers in tranches to the Bancroft Library of the University of Berkeley in California. A much larger portion is currently still held by a foundation established after the poet's death in July 2020 to administer her estate and is also scheduled to be sold to the Bancroft Library in the near future [as of December 2020]. This part of the papers will be referred to in the following as the "Albion Archives", named after the last residence of weiss, a house in the redwoods of the small village Albion on the Mendocino Coast in Northern California, 150 miles north of San Francisco.

The extent of other documents in the possession of private persons (e.g. fellow poets and other artists, friends and acquaintances who received manuscripts and letters from weiss) as well as in the archives of publishing houses and magazines cannot be specified at this time. In any case, it can be assumed with a high degree of probability that the poet produced copies of most of these works with a typewriter and archived them, which is why the majority of her oeuvre can be found in the Albion Archives. This is also true of the materials already archived in the Bancroft Library, which contain a large number of manuscripts of published and unpublished poems. However, these are mostly manuscripts written in colored (mostly silver) pencil on various types of colored paper, and are to be regarded as copies of the original typescripts, which weiss kept for herself. These typescript versions in the Albion Archives are almost certainly clean copies, as they contain hardly any corrections. Handwritten originals, sketches, notes or preliminary stages of the typescript versions were not mentioned by the author during her lifetime, and so far only a few such documents have been found in her papers, so that it can be assumed at this point that the poet did not consider it important to preserve them for posterity. weiss often emphasized that she did not make any revisions apart from corrections of obvious errors, so that the preservation of any earlier versions for text-genetic analysis would probably not have yielded much, although my 2021 documentary film *One More Step West Is the Sea* on the poet depicts a scene in which one sees how meticulously she could work on the stylistic improvement of a poem. However, in this particular case it is a poem of her last partner and percussionist Hal Davis that she is revising in collaboration with its author.

The work of weiss spans a period of over seven decades. Although she enjoyed telling the story that she wrote her first poem, called "es war einmal ein bär" ("once upon a time bear"), which has been published in the poetry collection *a fool's journey* (2012), at the age of five, and that she wrote a novel at twelve (which is not to be found in her papers and must be considered lost), the earliest surviving text – apart from "es war einmal ein bär" – dates from 1947 and was written during weiss' stay at a boarding school in Neuchâtel, Switzerland, while her father worked as an accountant for the U.S. Army Civilian Corps during the occupation of Germany in Frankfurt at the Office of Military Government headquarters, the task of which was the denazification and democratization of post-war Germany (and propaganda). It is a short untitled prose piece of half a page written on a typewriter, describing the landscape of the Swiss "Voralpen" (a part of the Western Alps). Despite its brevity, the text already contains elements that become characteristic of the poet's later work. The landscape is seen as a living organism, just like the desert in *Desert Journal* (1977): the Alps "laugh in a fierce way. Yet, in their wrinkles of stone and snow, they nestle and cuddle little velvet flowers—edelweiss—." Although it is prose, the rhythm and the repetition of similar sounds and syllables have the effect of echoing later poetry, in which, as here, colors also play a central role: "in the summer in the Voralps there's such green-ness and such blueness and such windness and such wildness … and such stillness."

Other prose texts from the late 1940s depict the attempts of short stories that were written, among other things, in the context of a "college contest" at Wright Jr. College in Chicago, after weiss had moved back from Europe to the United States in 1948. Among them is the story "No Credit", which begins with the detailed description of a certain Kenneth who finishes reading *Dr. Faustus* in his apartment and closes the book. In particular, the first references to the talented author's preoccupation with jazz, which can be found in this piece, is worth noting. Here's an excerpt:

> Seemingly concentrating on lighting his pipe, he divorced his mind from his action … and what do I want? … what do I need? With a sudden twist he coiled from the chair, flipped the radio to a compilation of trumpet and drums, and strode to the window. Neon lights wove glittering rainbows against the night, mocking the stars; one-dimensional buildings shamelessly bared their flat bodies into a monotonous perspective; stories below, beetle-cars sped the avenue—urbane, sophisticated, restless civilization … determined to boast, to advance, to plunge into … nowhere.

Such texts indicate that Ruth E. Weiss, as she signed her writings until 1952 (the "E." stands for her middle name Elisabeth), initially saw herself as a writer of a wider range of genres before concentrating on poetry. Rare poetological writings

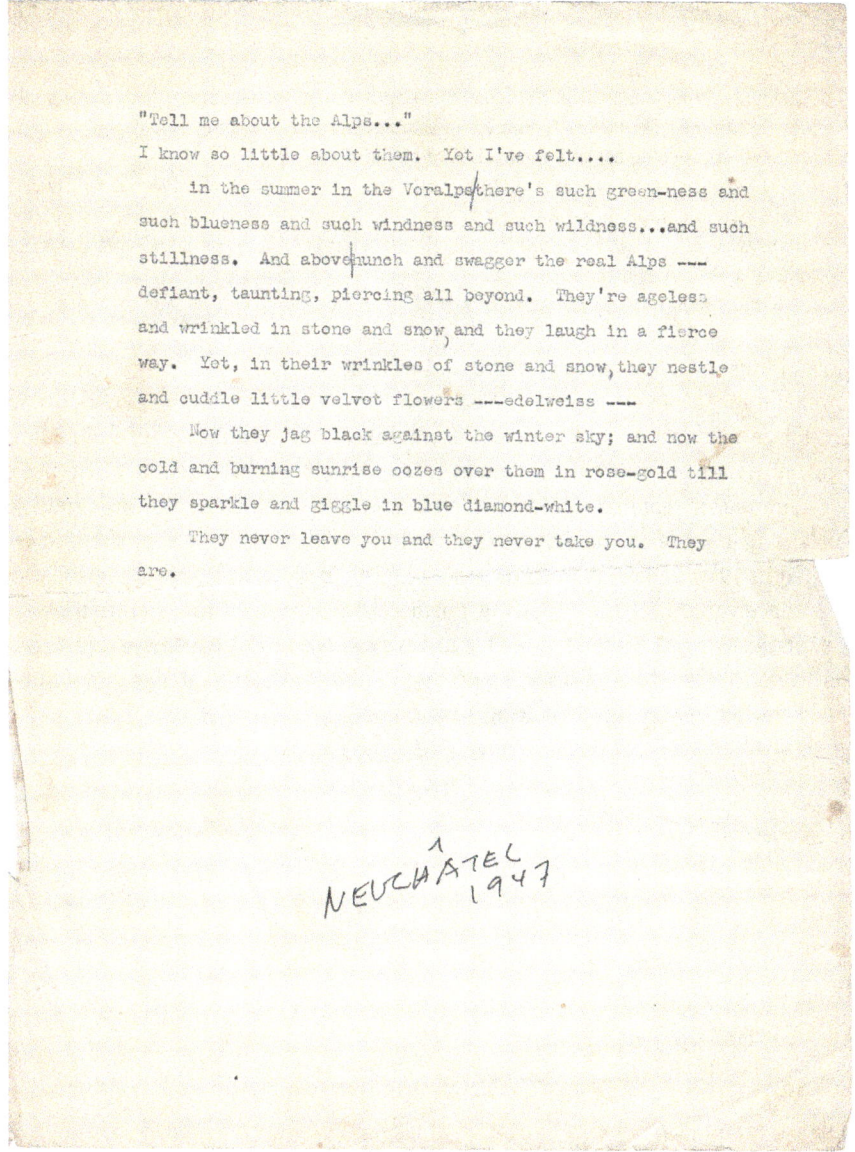

Fig. 1: "Tell me about the Alps ..." [incipit], manuscript, 1947

can also be found among these early works, which on the one hand refer to the intensive examination of certain authors, but on the other hand also reveal something about the young weiss' own writing style, which is still valid for her in later years. For example, the following untitled short text from around 1949 reads:

> Valéry mentions in *Monsieur Teste* that "the population of 'intellectuals'" consists of a rabble of puritans, speculators, believers, fools, anarchists, whores, priests, Caesares, martyrs, and even those who assume themselves to be children.
> I believe that an artist (I do not confuse artist and intellectual) is forced by his very perception to play all these. He must be all, to be himself—an itself that is always spectator. A spectator must *seem to be* participant. He must not only be those mentioned above; he must also be slave, mechanic, merchant—whatever and whomever the moment demands. As long as he doesn't lose himself in any of these effigies, he will be able to create.

Most of these early texts are undated and were archived by weiss in folders titled, for example, "1948 – 52 Chicago / New York / New Orleans", so the date of origin cannot be determined exactly (even if one takes into account the use of certain types of paper on which weiss wrote, typographical peculiarities and the like). For example, it remains unclear which works were created at the Art Circle in Chicago in 1949 or in the French Quarter in New Orleans, to where she had relocated in 1950 after a sojourn of some months in New York City. Nevertheless, it is conceivable that the following poem could have been the one that Ernest Alexander tore out of her typewriter in weiss' room at the Art Circle and urged her to perform in front of an audience, while the musicians on stage spontaneously started to play, thus creating Jazz & Poetry:

> Rhythm of the Bop ...
>   Gone notes ...
> Cool, cool ...
> Blow Prez blow
> I've got eyes.
> Solid.
> Daddie poke me.
> Bloodshot
> in stupored glaze.
> Up in highland.
> Blow Prez blow.

This poem, if not written in 1949, then at least before her arrival in San Francisco in 1952, not only testifies to the heavy influence of jazz already present in this early creative phase, but can also be regarded as Beat poetry *par excellence* with its rhythmic drive, spareness, and affinity for bebop – long before weiss met Jack Kerouac and the other Beats in San Francisco.

Other poems can almost certainly be dated to weiss' time in New Orleans in the early 1950s. Between August and November 1950, she published her first five texts in the weekly *The Old French Quarter News*, which was founded in 1941 and taken over by saxophonist Bruce Lippincott and Jeanne Smart in 1948. weiss' very first publication in the August 11, 1950 issue is an untitled short prose text, which, although it hints at a socio-critical leitmotif in the poet's work, is overly moralizing and negligible in content. The first paragraph: "They sip their beer and huddle together seeking response ... yet they find only the reflection of their own misery and loneliness ... These are the products of our age – of any age. The struggle of self against self, or self against environment is ageless." (weiss 1950)

In any case, the folder dated "1948–1952" contains a good two dozen unpublished poems that are typographically very similar to the typescripts of the *Old French Quarter News* publications and were written on the same type of paper, so there is little doubt that these were also created in the same period when she lived at 912 Toulouse in the bohemian part of the city in which she first dyed her hair green. Among them are, besides many poems still marked by adolescent *Weltschmerz*, which are all the more interesting for the biographer, since they often deal with motifs such as loneliness, fragility and insecurity, also small pearls like the following:

```
if Kant
     can't
FOLLOW FEELINGFUL ...
     the experience
of each moment
     in
     an
     is
THROUGH
     darkness ...
     nestling close
     the can't
TO
LIGHT
```

Between 1951 and 1958 there were no further publications by ruth weiss. Even after she had arrived in San Francisco in 1952 most of the manuscripts are not dated and it is difficult to make precise assertions about the actual date of origin. A folder with about 80 mostly unpublished poems is only labelled "Chi[cago] S[an] F[rancisco] 1952–60", another one contains 30 poems which, according to the label, were written between 1952 and 1966. In contrast to later published works of the same period such as "One More Step West Is the Sea" in weiss' first volume of poetry *Steps*, 1958, or "Single Out", written in 1958 and published in

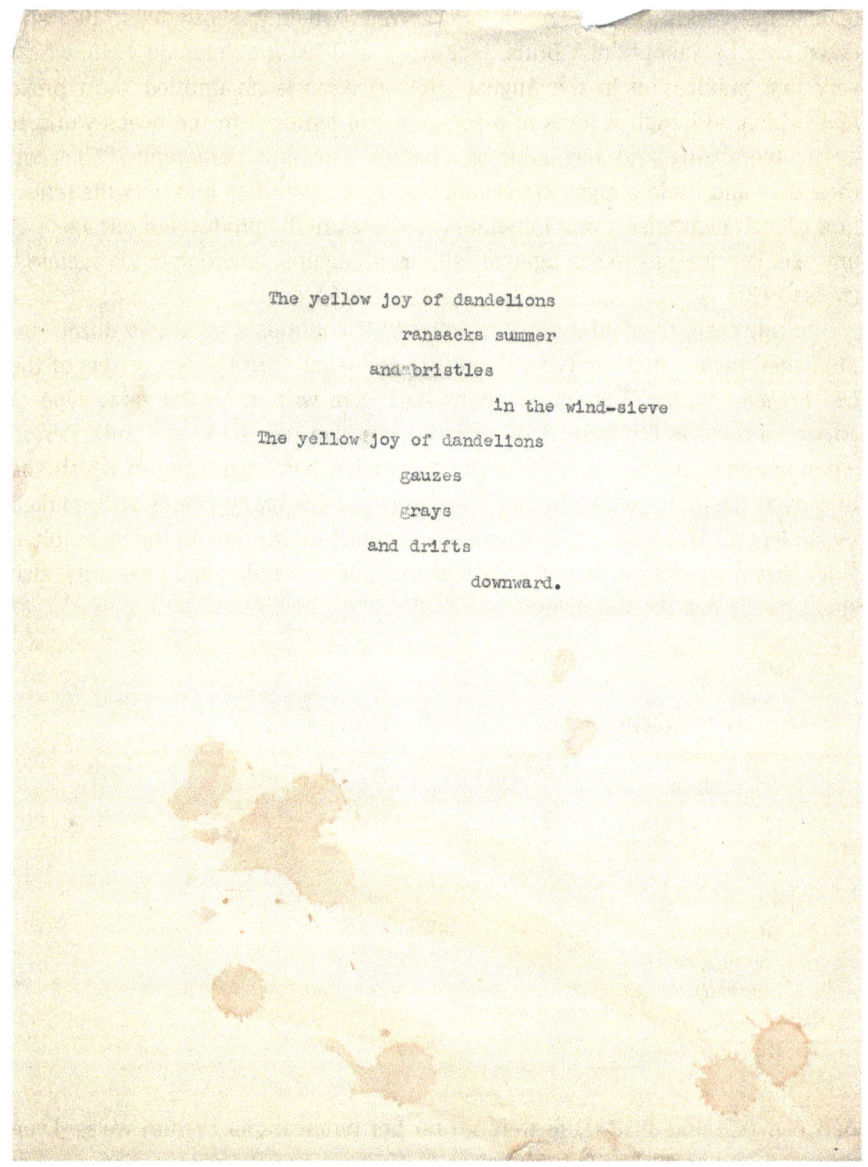

Fig. 2: "The yellow joy of dandelions" [incipit], manuscript, ca. 1948–1952

1978 in the poetry collection of the same name, there are also virtually no autobiographical traces to be found in the majority of these works, which could reveal something about the circumstances and motives for the production of the text. Only occasionally and in passing references do accounts of the life in North Beach appear, such as in "Four Small Eruptions At Vesuvio" with four short poems that sketch snapshots from the bar next to the City Lights bookstore on the corner of Broadway and Columbus:

> tear a sudden
> page
> the ball
> bounces between
> those who talk
> too well ...
> at a moment
> talk
> cannot—

Or an untitled work in which the poet records her pain over her first separation from Mel Weitsman (also known under the name of Sojun Roshi, who was the abbot of the Berkeley Zen Center until 2020, and after his passing in January 2021 Hakuryu Sojun), whom she was later to marry. Rather than autobiographical sketches, they are mostly filigree, seemingly lapidary observations of the surroundings, enlivened by the imaginative richness of the speaker:

> this is the wall
> that holds a cottage ...
> that is a castle,
> that holds the tower
> of the six storied virgins—
> when princes, pirates and wise men
> had perished ...
> an ant
> crawled
> through
> a crack.

The everyday observation thus opens a door to her inner world for a brief moment, which in turn elevates the object to something solemn due to intellectual vigour. One could perhaps call this process inter- and transpersonal self-transcendence. The term "stream of consciousness" used by weiss herself for her way of writing during this period (Antonic 2014, n.p.) is certainly not entirely accurate, as it is a fictional imitation of the thought process of, for example, a novel character. For later works, *automatic writing* and *free writing* techniques are probably more appropriate. The banal process of the ant crawling through the crack in the wall finally leads back to the here and now.

This form of poetry runs through the further work of ruth weiss and is also well-known in numerous publications. The simplicity is later condensed into the haiku form, with which she began to work in the 1950s. As is commonly known, weiss wrote numerous haiku together with Kerouac in 1955 during night-long sessions in her room at the Wentley Hotel, as a kind of dialogue between her and the author, who was hardly known at the time. weiss had always regretted that she did not keep any of these haiku and had no idea where they ended up. They are not to be found in the Jack Kerouac Papers either, so for better or worse they must be considered lost. It is remarkable, however, that there are no other haiku from this time in the weiss Papers. Closest to the form are a few four-line poems. The last line of each – similar to haiku – plays with the meaning of what was said before and makes "fall the tooth of that ruminant-of-the-absurd which the Zen apprentice must be, confronting his *koan*" (Barthes, 72):

> it's that silent flower
> that shouts so
> it could wake
> all the burning

The manuscripts of weiss' first volumes of poetry from this period (*Steps, Gallery of Women, South Pacific, Blue in Green*) are to be found in her papers as well, and so are some works published in later collections, such as "Chopsticks" (in *Single Out*) and the travelogue "Compass", unusual in form for weiss, which she wrote during her six-month journey with Mel Weitsman across Mexico (excerpts published in *Single Out*, complete version in *Can't Stop the Beat*, 2011). It is also worth mentioning that some poems of the *Gallery of Women* series are preserved that were not included in the poetry collection and probably reserved for a second volume. The thirteen manuscripts of the poems published in *Beatitude* in 1959 (see the bibliography in this volume) are also preserved in the Albion Archives.

In 1960, and even before the production of her film *The Brink*, whose text – which differs greatly from weiss' long poem of the same name published in *Single Out* – can also be found in the Albion Archives, the poet turned to the theatre and in a short period of time wrote ten plays, all of which are archived in the papers. In the volume *No Dancing Aloud* (2006), three of these are published ("No Dancing Aloud", "M & M" and "Figs"), as well as the later pieces "The Thirteenth Witch" (1981), "Torch-Song For Prometheus" (1999) and "One Knight and One Day" (2002). Some of these plays, written in the tradition of the Theatre of the Absurd and heavily influenced by Eugéne Ionesco's plays, as confirmed by the poet in a conversation (weiss 2017), have been performed over the decades on small stages, mainly in San Francisco, such as the "jazz play" "B Natural" (together with "M & M" at the Monday Blues Theatre of the Coffee Gallery in San Francisco, 1961, at the San Francisco Poetry Festival in the Spaghetti Factory and at St. Mark's Church in New York, both in 1973), in which the two contrary characters Bela, "dressed stark black tights, no jewelry," and Barta, "she glitters —dress, beads, etc., carrying a beaded bag and a bell," named after the composer Béla Bartók, meet during a party and enter into a playful dialogue with each other, which makes up the whole of the plotless play. Other unpublished pieces are entitled "Late One Early", "Media – Medium", "Misprints", "Shoes", "Two Players & Five Clubs", "The 61st Year to Heaven", the latter performed in 1961 at the Gate Theatre of Sausalito, and "Stop That Flower".

"Stop That Flower" is of interest because it is the original script of the underground film *The Flower Thief* by Ron Rice. weiss was asked to write a "poetic script" for this project (weiss 2017). In it, a certain Mrs. Mead from San Francisco, who was to be played in the film by weiss' friend and later Andy Warhol Superstar Taylor Mead, and the tourist Mr. Bees are waiting at three o'clock in the morning on the corner of Powell and Market St. for a cable car, which is not in operation at this time of day. Later, a Mr. Buzz and a Mrs. Noozeprint (a malapropism of "newsprint") join them, contributing nothing more to the absurd dialogue than quotes from daily newspapers. The six-page manuscript, like most of the weiss' plays without plot, runs into a dead end. At the very most, one could say that in the end the stranger Mr. Bees is accepted and welcomed into the group by the locals:

> MR. BEES: what we call foreign affairs
> is no longer foreign affairs
> it's a local affair
>
> MRS. MEAD: what relation has the YOU TOO
> to the internal situation
>
> MRS. NOOZEPRINT: the alleged summary of
> is absolutely untrue
>
> MRS. MEAD: THEY could then fly
> over any part of the territory
> and we would wave to them
> from below
>
> MR. BUZZ: the clown is on his way
>
> MR. BEES: watch from him
> on the best maintained
> all faith cemetery

ALL: AND THIS MORNING
                      WE ALL
                      ARE STILL ALIVE

The script was finally discarded by Rice in favor of a completely free improvisation in shooting and editing, where the soundtrack consists of randomly received radio programs. Taylor Mead is featured alongside Bob Kaufman and others in the final version of the film. A five-minute scene with ruth weiss was discarded for the final version, as Lars Movin also writes in his chapter on *The Brink* in this volume. The title *The Flower Thief* can be attributed to ruth weiss, at least according to her own statement (weiss 2017). And it was only after Rice saw no use for the script anymore that weiss decided to rewrite it into a play and give it the title *Stop That Flower*, as she didn't want it to be confused with the film and ultimately found the new title better, as she explained in a conversation (cf. weiss 2017).

Until now, it has been assumed that from 1960 onwards, weiss was involved in many projects outside poetry – including the production of *The Brink*, theater work, organizing reading series, appearing as an actress in Steven Arnold's films, etc. – which is why sixteen years passed between her fourth book *Blue in Green* from 1960 and her next book publication, *Light and Other Poems* (1976), even though she wrote her main work *Desert Journal* between 1966 and 1969, which she was not able to publish until 1977. An examination of her private archives, still in the presence of the poet in 2017 and 2018, however, brought to light the fact that weiss was also very productive as a poet between 1960 and 1976, publishing over fifty poems in 32 publications, including anthologies such as the *Beatitude Anthology* (1960), *Blueboy's Revenge* (1972), *Peace & Pieces: An Anthology of Contemporary American Poetry* (1973), *This is Women's Work: An Anthology of Prose and Poetry* (1974), *Contemporary Fiction: Today's Outstanding Writers* (1976), as well as litmags and magazines such as the New York *Beatitude* spin-off *Beatitude/east* (1961), *Film Culture* (NYC 1963/64), *Outburst* (London 1963), *The Human Eye* (LA 1967), *Peace & Pieces Review* (1976). Besides, she appeared regularly in local underground newspapers such as the *Haight Ashbury Tribune* in the 1970s or the legendary "erotic art newspaper" *Love Lights*, which was bought by many in the hope of finding more sex inside because of its covers "camouflaged" with soft pornographic subjects, but where "only" poetry, essays, comics, and the like were to be found. Not only the manuscripts of these works can be found in the Albion Archives, but also all the author's copies of the publications, many of them rare collectibles today, which seldom found their way into the poet's books at a later point.

The degree of extensive output did not decline in later decades. Rather, weiss expanded her range of publication possibilities; initially, after moving north to

Albion in 1983, in the guise of local media such as *The Pacific Coast News & Literary Review*, *A & E: Arts & Entertainment*, *New Settler* or *The Mendocino Beacon*, and after achieving greater recognition, also in international and relevant Beat publications such as *Kerouac Connection*, *Transit: A Little Magazine of the Beat Generation*, *Gargoyle*, *Sparring With Beatnik Ghosts*, or anthologies and magazines such as *Beat Attitude: Femmes Poètes de la Beat Generation* and *Action Poétique* in France, the Swedish *Lyrik vännen*, or *Austrian Beat* in Austria (see the bibliography in this book).

In addition, there are numerous unpublished manuscripts, including a large number of poems from around the mid-1990s onwards, which – similar to the poetry collections *Gallery of Women* and *For These Women of the Beat* – each focus on a person in the poet's circle, from which weiss later wanted to compile a volume of poetry entitled *Homage*, but was unable to complete due to her declining health. The occasion of most of these poems are birthdays or deaths, but sometimes also critical situations in which the subjects find themselves. They often possess the gesture of magical spells and are to be seen as invocations in the multiple sense of the word (e. g. as invocation of spirits and other entities or as an alternative to prayers). In terms of content, they often deal with relatively banal situations, which are nonetheless naturally important for the respective individuals, such as the separation of a friend in Mendocino from her ex-lover whom the poet wishes much strength ("For Sally Stewart", 2005), but also situations between life and death, written for colleagues and friends, for example for a 14-year-old boy while he was in a coma after being involved in a car accident ("For Jan at 14", 2017). Some poems can also be read as spells to avert unpleasant situations, such as when the poet tries to cope with her own difficult separation from her long-term partner Paul Blake in 2008, who suffered from psychological problems towards the end of his life, aggravated by alcohol and other drugs, or such as in "Easy Exit" (2017), which is about getting rid of a difficult tenant from her property:

> pack what is of value to you
> go to your next destination
> it is calling you
> to enter a new phase
> where the music you do
> will give you your due
>
> clean up your act
> it's a fact
> it will heal your wounds
> as you find your way
> on the golden path

Fig. 3: "hang in & let go" [incipit], broadside, 1994, illustration by Paul Blake

of your destiny
it may seem an impossible task
but all you have to do
is ask

Though the subjects of weiss' poems are rarely public persons the poet did not know personally, such poems can also be found. In the case of "re:ELECTION – fact or fiction" on the occasion of the US presidential elections in 2016, those can also take political dimensions:

DUMP TRUMP
his platform rotting wood
his band plays on
while the timber crash

while the timber crash
his band plays on
his patform rotting wood
DUMP TRUMP
(weiss 2020, 129)

This incantation, which was the last work published during her lifetime (as a facsimile in the Austrian literary magazine *Triëdere*), did not have the desired effect until 2020. Many of these works can be found in the collection housed at the Bancroft Library as well as in the Albion Archives. A few, like the one quoted above, have been printed in litmags and other journals.

The audiovisual archive of ruth weiss is also relevant. From 1958 to 1984, over 30 reel-to-reel tapes have been preserved, documenting various performances by the poet, both public and private. As has been mentioned previously (see, for example, the chapter on weiss' Jazz & Poetry by Hannes Höfer in the present volume), unfortunately, there are neither recordings of the performances in The Cellar from 1956 nor recordings that could document earlier Jazz & Poetry performances by weiss. The earliest recording in the Albion Archives is a reading with ruth weiss and Anne McKeever from 1958 without musical accompaniment, which is significant not only because of this collaboration, but also because there are hardly any other documents of McKeever's work. The poet and painter McKeever, who can be linked to the Beat movement and, like weiss, was a friend of Philip Lamantia, was living in Mexico at the time when weiss and Mel Weitsman visited the country, and later spent a few weeks in San Francisco in the summer of the same year, where she performed with weiss at the Labaudt Gallery on Gough Street. The poets performed their program *Tabulation of Troubadour*, a collaboration that improvised questions to the audience (e. g. "In how many peo-

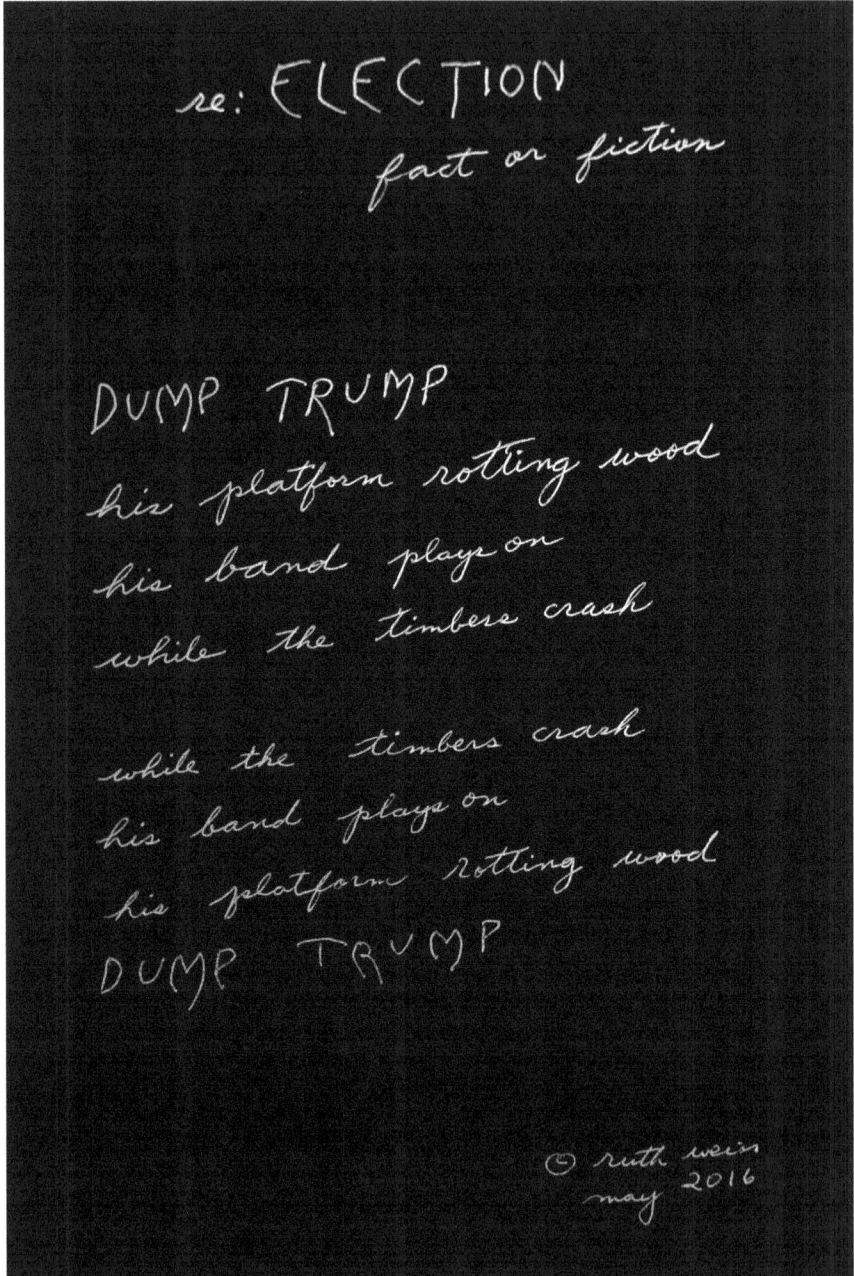

Fig. 4: "re: ELECTION: fact or fiction", manuscript, 2016

ple's dreams do you appear?") and produced spontaneous poems from the answers in dialogue, anticipating audience participation of later *Desert Jounal* perfomances.

Some of the recordings from the early 1960s are jam sessions with musicians recorded in the weiss/Weitsman house, while other tapes contain documents such as an audio recording of the premiere of weiss' play *The 61st Year to Heaven* from 1961 directed by Diane Varsi, who had ended her career as a Hollywood actress in 1959 and moved to San Francisco, or the original soundtrack of *The Brink*. Several tapes preserve sessions with bassist Benfaral Matthews, recorded in Topanga Canyon near Los Angeles in the late 1960s, where weiss lived with Paul Blake for two years, and contain several early performances of the then unpublished *Desert Journal*. There is also the 16-minute recording of a jam session with Bruce Lippincott in Santa Fe, New Mexico, from November 1969, who, just like weiss, had relocated from New Orleans to San Francisco and had become a member of the Cellar Quintet in the mid-1950s.

In addition, there are several recordings of weiss' radio performances among the tapes in which she recites poems and gives interviews such as Radio KQED 1970, Radio KPFA 1973, Radio KPOO 1973. The latter features a rare recitation of the complete long poem "Single Out" including a stunned commentary of the perplexed radio show host after the performance, which the record label Counter Culture Chronicles released on the cassette *ruth weiss – 2018 Live at the Beat Museum and Other Recordings from the Archive* in 2019. An important contemporary document is also a reading with Jack Micheline at the Blue Dolphine, San Francisco, in March 1973, where both poets should have read an equal amount of time, but after a while weiss decided to stop as she would rather listen to Micheline, as she announces. In addition, performances with the drag queen Mona Mandrake (a.k.a. Michael Shain) from 1973 and 1980 have been preserved, demonstrating that weiss was not only experienced as a solo reader and jazz & poetry performer, but also as a variety artist. Unusual during the performance of 1973 is the recitation of a poem by Erich Kästner in German ("Announcement of a Chansonette") and the chanson text "La Ville Inconnue" by Edith Piaf in French and weiss' translations into English ("Chansonette" and "In an Alien Town" respectively).

Towards the end of the 1990s, weiss' increasing popularity coincided with easier ways to make recordings thanks to digitalization, which is why a large number of performances both from Europe and the United States are preserved from that time forward. This also applies to video documentation, which hardly exists before the mid-1990s (despite the existence of some VHS tapes). In addition, in later years weiss became increasingly aware of the importance of documenting her artistic career and was keen to ensure that at least an audio recording of each performance was produced, if possible also a video. From the mid-1990s on, some important performances are documented on VHS videos, such as an interview with weiss at the San Francisco Public Library in 1994, stage talks at events of the Bancroft Library, at the De Young Museum San Francisco and at the San Francisco Book Fair 1996. Later, one can find on DVD-R numerous Jazz & Poetry performances in the Bay Area, in the Mendocino Region, in New York, New Orleans, Ohio, in Austria, Germany, as well as appearances on TV and in documentaries (e.g. on the Beat Generation or about Steven Arnold), the performance of three plays in Vienna in 2006, or weiss' legendary performance at the 40th anniversary of the "Summer of Love" in Golden Gate Park in 2007, which is analyzed by Peggy Pacini in the present volume.

The Albion Archives also contain originals of visual art works, such as the poet's haiku paintings, some of which were printed as facsimiles in the Austrian magazine *Flugschrift* (2017), as well as works of art by artist friends such as Madeline Gleason, Sutter Marin and weiss' companion Paul Blake, which are currently archived by the Calabi Gallery in Santa Rosa, CA. Secondary literature on ruth weiss and on Beat Poetry in general, an extensive library that can provide information about the poet's perusals, as well as life testimonies such as a school notebook of ten-year-old weiss from the school year 1937/38, in which the student had to record the annexation of Austria by Nazi Germany, numerous photographs, documents, correspondence, etc. complete the collection. However, at this stage it is not possible to say which of these works and materials will be taken over by the Bancroft Library and which will remain in private hands or be sold to other institutions.

This brief outline cannot provide more than a rough overview. A detailed picture of the ruth weiss Papers can only be provided by an itemized index, which has the scope of an entire book of its own and therefore cannot be printed in this publication. It is obvious that the ruth weiss Papers are of immense importance for future research, both in terms of studies of weiss' work and her biography. If one is involved in working on ruth weiss' life and career, one will not be able to avoid dealing with her papers, which contain hundreds of unpublished poems, as well as the hitherto over three hundred little known publications in litmags, anthologies, newspapers, etc., which enrich the thematic and formal spectrum of the poet's work, known from her book publications.

In addition, the collection also offers a wealth of material for analyses in the fields of the Beat Generation in general, Gender Studies, Theatre Studies, which, for example, refer to post-war American off-theatre, or Exile Studies. An edition of weiss' uncollected poetry as well as a selection of previously unpublished poems in book form would obviously be desired, although particular caution is called for here, as ruth weiss opted to keep many of her poems unpublished during her lifetime, and for good reasons. An edition of weiss' complete poetry, however, should of course include all the tangible works of the poet. In any case, such a project would be conceivable at least in digital form.

Works Cited

All of ruth weiss' unpublished works quoted in this chapter can be found in the Albion Archives of the ruth weiss Papers. Since a transfer of this collection to the Bancroft Library of the University of California at Berkeley is imminent [status April 2021, unless other decisions unknown to the author are going to be made] and no sorting and cataloguing has yet been carried out, these sources are not described in more detail here.

Antonic, Thomas. "Vienna never left my heart." Interview by Thomas Antonic. *European Beat Studies Network*, 2014. Web: https://ebsn.eu/scholarship/voices/ruth-weiss-interviewed-by-thomas-antonic/ Accessed 20 Oct. 2020.
Barthes, Roland. *Empire of Signs*. Transl. Richard Howard. New York: Hill and Wang, 1989.
weiss, ruth."They sip their beer." [Incipit] *The Old French Quarter News*, 11 August 1950, p. 5.
weiss, ruth. Personal interview by Thomas Antonic, August 9, 2017, Albion, California. Unpublished.
weiss, ruth. "re: Election – Fact or Fiction." *Triëdere*, vol. 11, no. 22, 2020, p. 129.

Image Sources

Img. 1–3: Scans from the ruth weiss Papers, courtesy of ruth weiss.
Img. 4: ruth weiss: "re: ELECTION: fact or fiction." Facsimile. Published in *Triëdere*, vol. 11, no. 22, 2020, p. 129.

Thomas Antonic, Paul Pechmann
ruth weiss: Complete Bibliography

This bibliography contains a complete list of ruth weiss' published works, both printed and audiovisual, excluding unofficial releases such as unauthorized reprints, online publications of her works, or YouTube videos. As mentioned in Thomas Antonic's chapter in this book, the large corpus of weiss' unpublished works will not be listed here. The editors made a conscious decision to list each publication in chronological order instead of conflating multiple publications of one poem under its first printing. A reference to the original publication is included with the listing of the republications. In this way, it is easier to maintain an overview of weiss' increasing output over the span of many years. The second part of the bibliography lists all publications of the first part in alphabetical order (ignoring any initial *A*, *An*, *The*, even when listing incipits, referencing the list in the first part). The third part is a selection of literature (in several languages) and audiovisual media about ruth weiss that the editors consider relevant. Texts in which weiss is only mentioned briefly are mostly not included. Likewise, countless portraits of the poet in newspapers, journals and online that repeatedly recount her story in a similar way and do not add any new information have been omitted. The bibliography largely follows MLA style – with the exception of ruth weiss' name, of course, and other obvious idiosyncrasies.

I Works by ruth weiss

I.1 Books

[1] *Steps*. San Francisco: Ellis Press, 1958.

[Content: "One *More* Step West Is the Sea"; "41 Dragon Steps"]

[2] *Gallery of Women*. San Francisco: Adler Press, 1959.

[Content: "A la Blake"; "Any"; "The Marvelous Mind That I Am"; "Ana in the English Lesson"; "Oh Ana Ana Oh"; "Ana in Mexico"; "Giulletta"; "To Sue"; "Dora"; "Still Dora"; "Tale about Dori or A Lioness With Mane"; "To Nancy Nefertete"; "Idell"; "Boo"; "Joan"; "Sueko"; "Kitty"; "To Lois and Her Cricket"; "To Gogo"; "To V W"; "Phyllis"; "To Phylis One January"; "Leslie"; "Eva"; "Sheila"; "Beverly"; "Cynthia"; "Sandy"; "Iva"; "Marianne"; "Shirley"; "Judy"; "Rita"; "June"; "Helen"; "Diane"; "Cockie with Earth-Paintings"; "To Dee or Robin's Nest"; "Pattie"; "Words No Words for Dance Fumi"; "Jean"; "Sor Juana"; "Laura"; "Pier"]

[3] *South Pacific*. San Francisco: Adler Press, 1959.

[Content: "District God"; "Ancestor Figure"; "Drum Base Figure"; "Painted Bark Cloth"; "Seat Shaped As Bird"; "Suspension Hook"; "Suspension Hook with Bird Form"; "Modelled Mask"; "Bowl with Carved Figures As Handles"; "Lime-Tube Stopper"; "Dagger Shaped as a Bird"; "'Skakabul' Mortuary Dance Object"; "Carving Use in Mortuary Ceremonies"; "Painted Decorative Carving"; "Decorated Bowl Shaped As Bird"; "Incised Club"; "Decorated Comb"; "'Musumu' Canoe Prow Figure"; "Turtle-Shaped Platter"; "Pig-Killer Adze-Shaped Club"; "'Tale' Carved Door Jamb"; "Belt"; "In A Japanese Tree Garden"]

[4] *Blue in Green*. San Francisco: Adler Press, 1960.

[Content: 12 untitled poems numbered "(1)" to "(12)", except "(8)" titled "lemonspiel" and "(12)" titled "john hoffman"; incipits: (1) "green thoughts"; (2) "ocean marriage"; (3) "rock the many"; (4) "crag"; (5) "seven stones"; (6) "ghost-scents"; (7) "salt-blue"; (9) "diane"; (10) "warm swarms the wind"; (11) "in a breezy wink"]

[5] *Light and Other Poems*. San Francisco: Peace & Pieces Foundation, 1976.

[Content: "Centered"; "Sharp Up"; "Oyster Stew"; "The Rock on the Road"; "Coyote"; "Vortex"; "Candle & Leaves" {= series of poems, including: "Sena"; "Us"; "Deirdre"; "Kaisik"; "All"; "Dean Antony"; "Ghia"; "Irene"; "Moss"; "Dedication"; "Tse-Wah-Te-Ay"; "Mutti"; "Paul"} "Light" {series of poems titled "One"; ""Two A"; "Two B"; "Two C"; "Three"; "Four A"; "Four B"; "Five A"; "Five B"; "Five C"; "Six A"; "Six B"; "Seven A"; "Seven B"}]

[6] *Desert Journal*. Boston: Good Gay Poets, 1977. [Reprint: New Orleans: Trembling Pillow Press, 2012.]

[Content: 40 poems titled and numbered as "First Day" until "Fortieth Day"]

[7] *Single Out*. Mill Valley: D'Aurora Press, 1978.

[Content: "June 23rd, 1977 – Anna Akhmatova"; "Single Out"; "Four"; "Compass (excerpts from)"; "The Brink"; "Chopsticks"; "Anaïs"; "Something Current"]

[8] *13 Haiku*. Mendocino: Attic Press, 1986.

[Content = selection of poetry from the series "All Numbers Work in Time", including: "4"; "84"; "10"; "14"; "713"; "1127"; "1327"; "4583"; "6283"; "8383"; "11684"; "21485"; "8186113"]

[9] *For These Women of the Beat*. San Francisco: 3300 Press, 1997.

[Content: "Foreword"; "Brenda Knight"; "Helen Adam"; "Jane Bowles"; "Ilse Klapper"; "Madeline Gleason"; "Josephine Miles"; "Joan Vollmer Adams Burroughs"; "Vickie Russell"; "Helen Hinkle"; "Carolyn Cassady"; "Luanne Henderson & Anne Murphy"; "Edie Parker Kerouac"; "Stella Sampas"; "Joan Haverty Kerouac"; "Gabrielle 'Mèmére' Kerouac"; "Eileen Kaufman"; "Mary Fabilli"; "Diane di Prima"; "Barbara Guest"; "Elise Cowen"; "Joyce Johnson"; "Hettie Jones";

"Billie Holiday"; "Joanne Kyger"; "Denise Levertov"; "Joanna McClure"; "Janine Pommy Vega"; "Elsie John"; "ruth weiss"; "Aya Tarlow"; "Mary Norbert Körte"; "Brenda Frazer"; "Lenore Kandel"; "Anne Waldman"; "Jan Kerouac"; "Natalie Jackson"; "Jay DeFeo"; "Gui de Angulo"; "Joan Brown"; "Author's Note"]

[10] *A New View of Matter / Nový pohled na věc*. Bilingual Czech/English. Transl. Pavla Jonsson. Prague: Maťa, 1999.

[Content: "Dedication"; "A New View of Matter"; "I Always Thought You Black (excerpts)"; "I Hear with Love"; "Confrontation"; "June 23rd, 1977 – Anna Akhmatova"; "Mother's Day 1997"; "Listen Papa"; "Seventeenth Day"; "Twenty-Third Day"; "Train Song for Jack Micheline"; "Beast – Be a Saint"; "Single Out"; "The Brink"; "M & M (A One Act Play)"; "Light (excerpt)"; "This Is Really Real"; "Speak for Yourself"; "Chi Chi Chi"; "Turnabout"; + Photos and Bibliography]

[11] *Africa*. Lysice: Edice Satlava, 1999. [Text in English language; copyright page in Czech, 50 copies, signed.] [Reprint in English with an additional foreword by r.w.: Berkeley: Bancroft Library, 2003.]

[Content: "Foreword"; "Africa"]

[12] *Full Circle / Ein Kreis vollendet sich*. Bilingual English/German. Transl. Christian Loidl. Vienna: Edition Exil, 2002. [2nd, revised edition 2012]

[Content: "Torch-Song for Prometheus"; "Single Out"; "Full Circle"; "ruth weiss Speaks to the Editor"; "Biography"; "Bibliography"]

[13] *White Is All Colors / Weiß ist alle Farben*. Bilingual English/German. Transl. Horst Spandler. Ottensheim: Edition Thanhäuser, 2004.

[Content: {all poems untitled with following incipits} "snowflakes"; "she knows what she knows"; "the stories are the milky way"; "the paper is blank"; "it is told"; "the baby is born"; "ships of the desert"; "the baby is always white"; "snowflakes" {2}; "a cave in the dark continent"; "peaks of snow"; "butterfly on blossom"; "white dove on the wing"; "the bells are ringing"; "a promise kept"; "white light & soft"; "the gate is open"; "the point become line"; "snowflakes" {3}]

[14] *No Dancing Aloud. Plays and Poetry / Lautes Tanzen nicht erlaubt. Stücke und Gedichte*. Bilingual English/German. Transl. Horst Spandler. Vienna: Edition Exil, 2006.

[Content: "Her Number Was Not Called"; "The Legacy of Prometheus"; "No Dancing Aloud"; "One Knight and One Day"; "M & M"; "Figs"; "The Thirteenth Witch"; Essay by Horst Spandler: "ruth weiss and the American Beat Movement of the '50s and '60s"; "Plays by ruth weiss – Performed 1961–2006" (list); "Biographies"; "Acknowledgement"]

[15] *Can't Stop the Beat: The Life and Words of a Beat Poet*. Studio City, CA [Los Angeles]: Divine Arts, 2011.

[Content: Introduction by Horst Spandler: "ruth weiss and the American Beat Movement of the '50s and '60s"; "Ten Ten"; "I Always Thought You Black"; Photos of the Poet; "Post-Card 1995"; "Compass"; About ruth weiss; Bibliography]

[16] *A Fool's Journey/Die Reise des Narren*. Bilingual English/German. Trans. Peter Ahorner and Eva Auterieth. Vienna: Edition Exil, 2012.

[Content: "Once upon a Time a Bear"; "The Children – Yes!"; "A Fool's Journey"; "Get a Life"; "Suicide Dreams"; "Beast – Be a Saint"; "Listen Papa"; "Mother's Day 1997"; "Altar-Piece"; "Fire Works"; "An Invitation"; "From Me to You"; "Beyond the Palace Walls"; "1967"; "Chi Chi Chi"; "I Hear with Love"; "Speak for Your Self"; "Light Works"; Biography]

[17] *A Parallel Planet of People and Places: Stories and Poems*. Bilingual English/German. Trans. Jürgen Schneider. Zirl: Edition Baes, 2012.

[Content: "For These Women of the Beat" {all 40 poems}; "In Hall in Tirol"; "Interview in Austria"; "Bypass Linz"; "5x7–5x12"; "Zimzum Is Three"; "Sutter Marin"; "Train-Song for Jack Micheline"; "Meeting Jack Micheline – Head On"; "For Philip Lamantia"; "I Am Calling You – Laura Ulewicz, Obit into Orbit"; "2009"; "2010"; "2011"; "2012"]

[18] *Einen Schritt weiter im Westen ist die See*. Bilingual English/German. Transl. Horst Spandler. Wenzendorf: Stadtlichter Presse, 2012. Heartbeat 20.

[Content: "One More Step West Is the Sea"; "Light"; "Seventh Day"; "Eleventh Day"; "Twelfth Day"; "Sixteenth Day"; "Seventeenth Day"; "Twenty-First Day"; "Twenty-Second Day"; "Twenty-Third Day"; "Thirty-First Day"; "Thirty-Fifth Day"; "Fortieth Day"; "Something Current"; "For Carol Bergé"; "Turnabout"; Afterword by Horst Spandler {= "ruth weiss and the American Beat Movement of the '50s and '60s" in German under the title "'Timing is what matters' – Der lange Weg von ruth weiss"]

[19] *The Snake Sez Yesssss / Die Schlange sagt jetzzzzzt*. Trans. Brigitte Jaufenthaler. Vienna: Edition Exil, 2013.

[Content: "2013"; "29 Days From Desert Journal" {"First Day"; "Second Day"; "Third Day"; "Fourth Day"; "Fifth Day"; "Sixth Day"; "Eighth Day"; "Ninth Day"; "Tenth Day"; "Thirteenth Day"; "Fourteenth Day"; "Fifteenth Day"; "Eighteenth Day"; "Nineteenth Day"; "Twentieth Day"; "Twenty-Fourth Day"; "Twenty-Fifth Day"; "Twenty-Sixth Day"; "Twenty-Seventh Day"; "Twenty-Eighth Day"; "Twenty-Ninth Day"; "Thirtieth Day"; "Thirty-Second Day"; "Thirty-Third Day"; "Thirty-Fourth Day"; "Thirty-Sixth Day"; "Thirty-Seventh Day"; "Thirty-Eighth Day"; "Thirty-Ninth Day"}; "Forty-First Day – The Return"; "At 85 in 2013"; Biography]

[20] *Light / Lumière*. Bilingual English/French. Marseilles: Stereo Editions, 2017. (French edition of *Light and Other Poems*)

[21] *Flugschrift* 25. Ed. Thomas Antonic & Dieter Sperl. Vienna: Literaturhaus Wien, 2018. (incl. Audio CD: *Jazz & Haiku* [see I.3 Audio Releases])

[Content: Haiku paintings from the series "A Fool's Journey" and "Banzai"; incipits, recto from top left to bottom right: "the dance in the flame"; "one foot off the cliff"; "enough words! start brush"; "making room for love"; "there – there"; "air air air air air"; "in search for her word"; "is this really dark"; verso: "THE SEA BRINGS MESSAGE"; "those hives of the past"; "green green green the hill"; "a mask for each mood"; "the rose chose the poet"]

I.2 Publications in journals, magazines, anthologies, etc.

1950

[22] "They sip their beer" [Incipit]. *The Old French Quarter News* (New Orleans) 11 Aug. 1950: 5.
[23] "Her presence assailed the room" [Incipit]. *The Old French Quarter News* (New Orleans), 18 Aug. 1950: 5.
[24] "Loose-flapped in cloaked dignity" [Incipit]. *The Old French Quarter News* (New Orleans) 22 Sept. 1950: 4.
[25] "A broken stairway suspended" [Incipit]. *The Old French Quarter News* (New Orleans) 20 Oct. 1950: 5.
[26] "The mists swam across the night" [Incipit]. *The Old French Quarter News* (New Orleans) 3 Nov. 1950: 8.

1959

[27] "Two Poems." [Incipits: "the animals have come down" / "sound moves solids"] *Beatitude* 2 (1959): n.p.
[28] "Big Sur." *Beatitude* 3 (1959): n.p.
[29] "Seven Times Yelapa." *Beatitude* 4 (1959): n.p.
[30] "Exodus –– 1136." *Beatitude* 4 (1959): n.p.
[31] "Somewhere Else." *Beatitude* 4 (1959): n.p.
[32] "Maple-Bridge Night-Mooring." *Beatitude* 5 (1959): n.p.
[33] "Two Poems." [Incipits: "the tooth of the tiger" / "yellow-sweet"] *Beatitude* 6 (1959): n.p.
[34] "ocean-marriage" [Incipit]. *Beatitude* 7 (1959): n.p. [First publication see ref. # 4]
[35] "Ting A Ling" [Incipit]. *Beatitude* 8 (1959): n.p.
[36] "Yee Jun" [Incipit]. *Beatitude* 8 (1959): n.p.
[37] "Pain Ting" [Incipit]. *Beatitude* 8 (1959): n.p.
[38] "cerberus" [Incipit]. *Beatitude* 11 (1959): n.p.
[39] "It's a rape-night" [Incipit]. *Beatitude* 11 (1959): n.p.
[40] "From Compass." *Semina* 5 (1959): n.p.

1960

[41] "Poem." [Incipit: "the animals have come down"] *Beatitude Anthology*. San Francisco: City Lights, 1960. 87. [First publication see ref. # 27]

1961

[42] "Africa." *Beatitude/east* (New York) 17 (1961): 18–19.
[43] "Overheard: Coffeeshop Talk." *Beatitude/east* (New York) 17 (1961): 19.
[44] "Zimzum Is Three." *Sun* (San Francisco) 3 (1961): 19–20.

1962

[45] "who will throw the pie in eye?" [Incipit]. *Sun* (San Francisco) 7: 1962. 19.
[46] "The Blues Singer." (Translation and Adaption ruth weiss from the German by Erich Kastner [sic]). *Sun* (San Francisco) 7 (1962): 19.
[47] "Flower in Rock (after a series of paintings by Sutter Marin)." *Poetry Score* (Carmel, CA) 2 (1962): 13–15.
[48] "Karen Blixen." *Poetry Score* (Carmel, CA) 2 (1962): 15.
[49] "Sun Burnt to Kites." *Poetry Score* (Carmel, CA) 2 (1962): 16.
[50] "To Each Other." *Poetry Score* (Carmel, CA) 2 (1962): 17.
[51] "Yet it Is." *Poetry Score* (Carmel, CA) 2 (1962): 18.
[52] "Maple Bridge." *Poetry Score* (Carmel, CA) 2 (1962), 19. [First publication see ref. # 32]
[53] "For T. R. One Hot Noon for Reading." *Poetry Score* (Carmel, CA) 2 (1962): 20.

1963

[54] "Untitled." [Incipit: "to crack visions"] *The Galley Sail Review. A Poetry Magazine* (San Francisco) 9 (1963): 23.
[55] "Four Poems by ruth weiss" ["L' Avventura", "La Dolce Vita", "Rashomon", "The White Dove (After a Film by Frantisek Vlacil)"], *Film Culture* (New York) 31 (1963–64): 43.
[56] "The White Dove (After a Film by Frantisek Vlacil)." *Outburst* (London) 2 (1963): n.p. [First publication see ref. # 55]
[57] "Rashomon." *Outburst* (London) 2 (1963): n.p. [First publication see ref. # 55]
[58] "Guardian Angel." $M–C^2$ (Healdsburg) 3 (1963): n.p.
[59] "The Garden Within." $M–C^2$ (Healdsburg) 3 (1963): n.p.

1965

[60] "Fairy Tale." *Haight Ashbury Tribune* (San Francisco) 1.5 (1965): 14.
[61] "will sing & soar" [Incipit]. *Fux Magascean!* [spelled "Magascene" on the cover] (San Francisco) 1965: n.p.
[62] "Scraps a Vertical and Horizontal." *Fux Magascean!* [spelled "Magascene" on the cover] (San Francisco) 1965: n.p.
[63] "someone sensitive" [Incipit]. *Kiroku to geijutsu* (Japan) 12 (1965): 39.
[64] "we burn from the moment born" [Incipit]. *Kiroku to geijutsu* (Japan) 12 (1965): 49.
[65] "Shibu Wabi Sabi Furu." *Kiroku to geijutsu* (Japan) 12 (1965): 42.

1966

[66] "Sungtim." *Illuminations* (San Francisco) 2 [1966]: n.p.
[67] "A Trip." *Nightshade* (San Francisco) 1.6 (1966): 7.
[68] "Who? Me?" *Nightshade* (San Francisco) 1.6 (1966): 7.

1967

[69] "Water and Fire." *The Human Eye* (Venice, CA) (1967): 7.
[70] "if the spiderweb" [Incipit]. *The Human Eye* (Venice, CA) (1967): 7.
[71] "one could say" [Incipit]. *The Human Eye* (Venice, CA) (1967): 7.

1968

[72] "Every Suicide Is Fratricide. Nine Fragments for Kitty." *Miscellaneous Man: The New Los Angeles Quarterly of Literature & Art* 1 (Summer 1968): 49.

1970

[73] "here harry" [Incipit]. *The Funny New Shape* (San Geronimo, CA) 1970: n.p.
[74] "The Brink." *Matrix: For She of the New Aeon* (San Francisco) 1 (1970): 22.

1971

[75] "Single Out." *Matrix: For She of the New Aeon* (San Francisco) 2 (1971): 17–20.
[76] "Vita-Sheet." *Mark in Time: Portraits & Poetry / San Francisco*. Ed. Nick Harvey. San Francisco: Glide Publications, 1971. 62.

1972

[77] "The Blind Voyage." *Gypsy Table* (San Francisco) 2 (1972): 45–48.
[78] "ezra talked to me last night" [Incipit]. *Love Lights* (San Francisco) (1972): n.p.
[79] "Fuck Me or Graffiti." *Blueboy's Revenge*. San Francisco: The Littlefield Press, 1972. 72.
[80] "Yellow Gray and Black." *Blueboy's Revenge*. San Francisco: The Littlefield Press, 1972. 73.
[81] "Horizon." *Blueboy's Revenge*. San Francisco: The Littlefield Press, 1972. 74.
[82] "The Return." *Blueboy's Revenge*. San Francisco: The Littlefield Press, 1972. 75.

1973

[83] "Current Events or A Political Poem or I Hope This Is Not True Tomorrow." *Lewt* (San Francisco) (1973): n.p.
[84] "Requiem for Brew Moore." *Lewt* (San Francisco) (1973): n.p
[85] "Ray." *Peace & Pieces: An Anthology of Contemporary American Poetry*. Ed. Maurice Custodio, Grace Harwood, David Hoag, and Todd S.J. Lawson. San Francisco: Peace & Pieces 1973. 172.
[86] "Eleventh Day." *Panjandrum Poetry* (San Francisco) 2–3 (1973): n.p.
[87] "Miracle Burn." *One-Eighty-Five*. Ed. Alix. San Francisco: Mongrel Press, 1973. 21–24.

1974

[88] "Desert Journal. Second Day / Sixteenth Day." *This is Women's Work. An Anthology of Prose and Poetry*. Ed. Susan Efros. San Francisco: Panjandrum Press, 1974. 58.
[89] "Current Events or A Political Poem or I Hope This Is Not True Tomorrow." *San Francisco Phoenix* 2.10 (1974): 10. [First publication see ref. # 83]
[90] "Candle & Leaves #5." *Poet's Gallery* (San Francisco) 3 (1974): 9. [Published under the title "All" in *Light and Other Poems*.]

1975

[91] "Ghia." *Second Spring: The Magazine of the Adult Society* (San Francisco) April/May 1975: 19.
[92] "Us." *Second Spring: The Magazine of the Adult Society* (San Francisco) April/May 1975: 19.
[93] "Coyote." *Love Lights* (San Francisco) Spring 1975 [special issue]: n.p.
[94] "Centered." *Love Lights* (San Francisco) Spring 1975 [special issue]: n.p.
[95] "who's going to read your books?" [Incipit] *Love Lights* (San Francisco) Spring 1975 [special issue]: n.p.
[96] "Us." *Love Lights* (San Francisco) Spring 1975 [special issue]: n.p. [First publication see ref. #92]
[97] "And What About Money?" *Love Lights* (San Francisco) Spring 1975 [special issue]: n.p.

[98] "For Jimmie Lowe Last Night at That Pub in Jackson Place in San Francisco." *Love Lights* Spring 1975 [special issue]: n.p.

1976

[99] "To a Poem Called Mary Stagliano." *Peace & Pieces Review* (San Francisco) 2.2 (1976): 17.
[100] "say everything you mean" [Incipit]. *Peace & Pieces Review* (San Francisco) 2.2 (1976): 18.
[101] "Oyster Stew." *Peace & Pieces Review* (San Francisco) 2.2 (1976): 19.
[102] "delicate bird" [Incipit], *Peace & Pieces Review* (San Francisco) 2.2 (1976): 20.
[103] "Desert Journal. Tenth Day." *Room: A Women's Literary Journal* (San Francisco) 1.1 (1976): 48–50.
[104] "No Dancing Aloud." *Contemporary Fiction: Today's Outstanding Writers*. Ed. Maurice Custodio. San Francisco: Peace & Pieces Foundation, 1976. 98–114.
[105] "Between." *Boston Gay Review* 1 (1976): 12.

1977

[106] "George Abend." *Beatitudes – Mayday 77* (San Francisco) 27 (1977): n.p.
[107] "Billie." *Contemporary Women Poets. An Anthology of California Poets*. Ed. Jennifer McDowell, and M. Loventhal. San Jose: Merlin, 1977. 92.
[108] "Spirit-Talk." *Anthology of the First Annual Women's Poetry Festival of San Francisco*. Ed. by Todd S. J. Lawson, and William J. Lawson. San Francisco: The New World Press Collective, 1977. N.p.
[109] "A Point of Reference." *Sheaf* (San Francisco) 1.1 (1977): 11.

1978

[110] "An Untitled Story." *Second Coming* (San Francisco) 6.1 (1978): p. 63.
[111] "Sena." *19 + 1: An Anthology of San Francisco Poetry*. Ed. A.D. Winans. San Francisco: Second Coming Press, 1978. 58.
[112] "Dean Antony." *19 + 1: An Anthology of San Francisco Poetry*. Ed. A.D. Winans. San Francisco: Second Coming Press, 1978. 59. [First publication see ref. # 5]
[113] "Tse-Wah-Te-Ay." *19 + 1: An Anthology of San Francisco Poetry*. Ed. A.D. Winans. San Francisco: Second Coming Press, 1978. 60. [First publication see ref. # 5]

1979

[114] "J." [French, translated by Jack Hirschman] *Amerus: An International Lyripolitical Journal of Poetry and Graphics* (San Francisco) 1 (1979): n.p.
[115] "For Benfaral Matthews. Died October 29th 1975." *The Hollow Spring Review* (Berkshire, MS) 3.1 (1979): n.p.
[116] "Circle V Motel." *The Hollow Spring Review* (Berkshire, MS) 3.1 (1979): n.p.
[117] "Africa." *Maybe Mombasa* (San Francisco) 3 (1979): 4–10. [First publication see ref. # 42]
[118] "Something Current." *Maybe Mombasa* (San Francisco) 3 (1979): 10. [First publication see ref. # 7]
[119] "For Madeline Gleason." *Poetry Flash* (Oakland, CA) 75 (1979): n.p.

1980

[120] "Fifty-One." *Soup* (San Francisco) (1980): 60.
[121] "Altar-Piece." *Soup* (San Francisco) (1980): 61.
[122] "For Otter and Sutter." *Sheaf* (San Francisco) 1.3 (1980): 20.

1982

[123] "20." *Newsletter* (Mendocino) 4 (1982): n.p.
[124] "24." *Newsletter* (Mendocino) 4 (1982): n.p.
[125] "internal weather" [Incipit]. *Newsletter* (Mendocino) 5 (1982): n.p.
[126] "23." *Newsletter* (Mendocino) 6 (1982): n.p.
[127] "crane dance step by step" [Incipit]. *Newsletter* (Mendocino) 6 (1982): n.p.

1983

[128] "Ray." *Newsletter* (Mendocino) May 1983: n.p. [First publication see ref. # 85]
[129] "104." *Newsletter* (Mendocino) May 1983: n.p.
[130] "24." *Newsletter* (Mendocino) 22 (1983): n.p. [First publication see ref. # 124]
[131] "1984." *Newsletter* (Mendocino) 25 (1983): 1.
[132] "1223." *Newsletter* (Mendocino) 25 (1983): 2.
[133] "Fortieth Day." *A & E: Arts & Entertainment Magazine* (Mendocino) May 1983: 10. [First publication see ref. # 6]
[134] "16 – 17." *Ridge Review* (Mendocino) 3.2 (1983): 26.
[135] "31." *Ridge Review* (Mendocino) 3.2 (1983): 26.

1984

[136] "12 19 84." *Newsletter* (Mendocino) 26 (1984): n.p.
[137] "Sena." *Second Coming Anthology: Ten Years in Retrospect*. San Francisco: Second Coming Press [= *Second Coming* 11.1– 2], 1984: 217. [First publication see ref. # 111]
[138] "Coyote." *Second Coming Anthology: Ten Years in Retrospect*. San Francisco: Second Coming Press [= *Second Coming* 11.1– 2], 1984: 218. [First publication see ref. # 93]
[139] "31." *Mendocino Commentary* 226 (1984): n.p. [ruth weiss misspelled as "ruth wein"] [First publication see ref. # 131]
[140] "Infinite." *A & E: Arts & Entertainment* (Mendocino) March 1984: 20.
[141] "51083." *A & E: Arts & Entertainment* (Mendocino) March 1984: 20.
[142] "a balancing act" [Incipit]. *A & E: Arts & Entertainment* (Mendocino) April 1984: 33.
[143] "walking a tightrope" [Incipit]. *A & E: Arts & Entertainment* (Mendocino) April 1984: 33.
[144] "4." *A & E: Arts & Entertainment* (Mendocino) April 1984: 34.
[145] "20." *Mendocino Commentary* 230 (1984): 19. [First publication see ref. # 123]
[146] "1112." *Mendocino Commentary* 230 (1984): 19.
[147] "484." *Big River News* 13.5 [issue no. 188] (1984) [= Literary Issue 1984]: 24.
[148] "SB 883." *Big River News* 13.5 [issue no. 188] (1984) [= Literary Issue 1984]: 24.
[149] "1416 – 9." *Big River News* 13.5 [issue no. 188] (1984) [= Literary Issue 1984]: 24.
[150] "21684." *Big River News* 13:.5 [issue no. 188] (1984) [= Literary Issue 1984]: 24.
[151] "120." *Big River News* 13.5 [issue no. 188] (1984) [= Literary Issue 1984]: 24.
[152] "84." *Big River News* 13.5 [issue no. 188] (1984) [= Literary Issue 1984]: 25.
[153] "414." *Big River News* 13.5 [issue no. 188] (1984) [= Literary Issue 1984]: 25.
[154] "1223." *Big River News* 13.5 [issue no. 188] (1984) [= Literary Issue 1984]: 25. [First publication see ref. # 132]
[155] "21884." *Big River News* 13.5 [issue no. 188] (1984) [= Literary Issue 1984]: 25.
[156] "18." *Big River News* 13.5 [issue no. 188] (1984) [= Literary Issue 1984]: 25.
[157] "3584." *Big River News* 13.5 [issue no. 188] (1984) [= Literary Issue 1984]: 25.
[158] "47." *Big River News* 13.5 [issue no. 188] (1984) [= Literary Issue 1984]: 25.
[159] "Competition." *Big River News* 13.5 [issue no. 188] (1984) [= Literary Issue 1984]: 25.

[160] "Voicing." *A & E: Arts & Entertainment* (Mendocino) August 1984: 15–16.
[161] "For Benfaral Matthews." *A & E: Arts & Entertainment* (Mendocino) August 1984: 15. [First publication see ref. # 115]
[162] "Epilogue: Message from Ben." *A & E: Arts & Entertainment* (Mendocino) August 1984: 33.
[163] "Eighth Day." *A & E: Arts & Entertainment* (Mendocino) Nov. 1984: 40–41. [First publication see ref. # 6]

1985
[164] "985." *Mendocino Commentary* 268 (1985): 15.
[165] "985." *Anderson Valley Advertiser* 29.47 (1985): 2. [First publication see ref. # 165]
[166] "3584." *Beatitude 33: Silver Anniversary*. Ed. Jeffrey Grossman. San Francisco: Beatitude, 1985. 164. [First publication see ref. # 157]
[167] "4." *Beatitude 33: Silver Anniversary*. Ed. Jeffrey Grossman. San Francisco: Beatitude, 1985. 164. [First publication see ref. # 144]
[168] "21884." *Beatitude 33: Silver Anniversary*. Ed. Jeffrey Grossman. San Francisco: Beatitude, 1985. 164. [First publication see ref. # 155]
[169] "1416–9." *Beatitude 33: Silver Anniversary*. Ed. Jeffrey Grossman. San Francisco: Beatitude, 1985. 164. [First publication see ref. # 149]
[170] "5." *Beatitude 33: Silver Anniversary*. Ed. Jeffrey Grossman. San Francisco: Beatitude, 1985. 164.
[171] "Infinite." *Beatitude 33: Silver Anniversary*. Ed. Jeffrey Grossman. San Francisco: Beatitude, 1985. 164. [First publication see ref. # 140]

1986
[172] "41684." *Mendocino Commentary* 274 (1986): 19.
[173] "apple of god's eye" [Incipit]. *Drift Dodger* 4.2 (1986): 8.
[174] "The Light Rose Pink." *Mendocino Commentary* 290 (1986): 16.

1987
[175] "Terry O'Flaherty (Crashed 2/7/87)." *Mendocino Commentary* 296 (1987): 15.
[176] "Crashed 2/7/87." [i. e. "Terry O'Flaherty (Crashed 2/7/87)"] *Arts & Entertainment* March 1987: 30. [First publication see ref. # 175]
[177] "Four for Sutter." *Beatitude* 34 (1987): n.p.
[178] "the score of the trees" [Incipit]. *Mendocino Commentary* 309 (1987): 15.
[179] "Sylvia Coddington." *Mendocino Commentary* 315 (1987): 15.
[180] "Atone at One." *Noh Quarter* 3.1 (1987/88): 68.

1988
[181] "the score of the trees" [Incipit]. *Mirrors* (Gualala, CA) 1.1 (Spring 1988): n.p. [First publication see ref. # 178]
[182] "1988." *Mirrors* (Gualala, CA) 1.2 (Sommer 1988): n.p.
[183] "62088." The Creative Discourse 3.3 (1988): 2.
[184] "a new view of matter" [Incipit]. *Mendocino Commentary* 338 (1988): 15.

1989
[185] "5 x 7—5 x 12." *The New Settler* 38 (1989) [Literary Issue]: 28–29.

[186] "For Bobby Kaufman." *Would You Wear My Eyes? A Tribute to Bob Kaufman*. Ed. Jack Hirschman. San Francisco: The Bob Kaufman Collective, 1989. 53.
[187] "Easy Access (For Buckhorn Cove)." *The New Settler* 40 (1989): 42.
[188] "Sungtim." *The Illuminations Reader: Art & Writing from Illuminations, Pulse & Gar: Anthology of Contemporary Literature, Ecology & Politics*. Ed. Norman Moser [place and publisher unknown], 1989. N.p. [First publication see ref. # 66]

1990

[189] "Billy Cannon." *Mendocino Coast Jazz Society* [newsletter] Nov. 1990: n.p.
[190] "Afterfact." *The Mendocino Review* 9 (1990): 39.

1991

[191] "a zone around earth" [Incipit]. Moments In The Journey: 1991 Datebook: A Collective Effort by Local Artists for the Benefit of the Santa Cruz AIDS Project. [USA]: n.p., 1991. N.p.
[192] "1991." [a.k.a. "'rribet' to the rain" {Incipit}] Mendocino Coast Jazz Society [newsletter] Jan. 1991: n.p.
[193] "1991." [a.k.a. "'rribet' to the rain" {Incipit}] Mirrors (Gualala, CA) Winter 1991: n.p. [First publication see ref. # 192]
[194] "63." The Pacific Coast News & Literary Review 4.13 (1991): 6.
[195] "For Health of Ma Earth" [Incipit]. The Pacific Coast News & Literary Review 4.17 (1991): 7.
[196] "Beast – Be a Saint." The Pacific Coast News & Literary Review 4.17 (1991): 7.
[197] "Now, We Are Talking about the End of 1971" [incipit]. Minnie's Can-Do Club: Memories of Fillmore Street. San Francisco: Ayers & Company, 1991. N.p.
[198] "the claw of the beast" [Incipit]. Minnie's Can-Do Club: Memories of Fillmore Street. San Francisco: Ayers & Company, 1991. N.p.
[199] "Us." Minnie's Can-Do Club: Memories of Fillmore Street. San Francisco: Ayers & Company, 1991. N.p. [First publication see ref. # 96]
[200] "Turnabout." Am Here (Philo, CA) July 1991: n.p.
[201] "autumn leaves falling" [incipit]. A & E: Arts & Entertainment Nov., 1991: 17.
[202] "against all the odds" [incipit]. A & E: Arts & Entertainment Nov., 1991: 17.
[203] "cloud calling sisters" [incipit]. A & E: Arts & Entertainment Nov., 1991: 17.
[204] "Hear Us." Mendocino Commentary (Fort Bragg, CA) 7 Nov. 1991: 11.
[205] "A Ballad for Our Time." The Pacific Coast News & Literary Review 4.22 (1991): 7.
[206] "Hear Us." The Pacific Coast News & Literary Review 4.22 (1991): 7. [First publication see ref. # 204]
[207] "Flamenco for Harriet and Larry." The Pacific Coast News & Literary Review 4.22 (1991): 7.
[208] "Why Did You Go to Las Vegas to Die Ron Towe." The Pacific Coast News & Literary Review 4.22 (1991): 7.
[209] "This Is Really Real." The Pacific Coast News & Literary Review 4.22 (1991): 7.

1992

[210] "A Ballad for Our Time." *The New Settler* 65 (1992) [Literary Issue]: 49. [First publication see ref. # 205]
[211] "Hear Us." *The New Settler* 65 (1992) [Literary Issue]: 49. [First publication see ref. # 204]
[212] "Orion: Toxic Waste." *The New Settler* 65 (1992) [Literary Issue]: p. 49.
[213] "Litany for Wanda." *The Mendocino Beacon* 19 Mar. 1992: 4.

[214] "... why I write poetry" [incipit]. *The Gab: Monthly Guide to the Redwood Coast* March 1992, 17.
[215] "A View from Albion." *The Mendocino Country Environmentalist* May 21, 1992: 5.
[216] "A View from Albion." *Mendocino Coast Jazz Society* [newsletter] June 1992: n.p. [First publication see ref. # 215]
[217] "Paul Zipp." *A & E: Arts & Entertainment* July 1991: 28.
[218] "A View from Albion." *The New Settler* 68 (1992): 2. [First publication see ref. # 215]
[219] "Bomkauf You Did It Again." *Mendocino County Outlook* 2 (1992): 10.
[220] "Don Murray (for Ruth Murray)." *Mendocino County Outlook* 7 (1992): 12.
[221] "Billy Cannon." *In the West of Ireland: A Literary Celebration in Contemporary Poetry*. Ed. Martin Enright. Lisselton: Enright House, 1992. 69. [First publication see ref. # 189]

1993

[222] "if i were the rooster" [Incipit]. *Mendocino County Outlook* 12 (1993): 14.
[223] "red-golden his crown" [Incipit]. *Mendocino County Outlook* 12 (1993): 14.
[224] "Coyote." *Mendocino County Outlook* 18 (1993): 14. [First publication see ref. # 93]
[225] "Out with without a Home." *Homeless Times* 1.1 (1993): 7.
[226] "Bomkauf You Did It Again." *Bouillabaisse 3*. New Hope: Alpha Beat Press, 1993. N.p. [First publication see ref. # 219]
[227] "Paul Zipp." *The Tule Review* Fall/Winter 1993/94: 13. [First publication see ref. # 217]
[228] "A View from Albion." *The Tule Review* Fall/Winter 1993/94: 13. [First publication see ref. # 215]
[229] "Altar-Piece." *The Tule Review* Fall/Winter 1993/94: 13. [First publication see ref. # 121].

1994

[230] "Ma Earth." *OutLook for Kids* (May 1994): n.p.
[231] "Flame." *Women of Mendocino Bay* 4.1 (1994): 14.
[232] "Beast – Be a Saint." *Poetry at the 33: An Anthology*. Ed. Lee Hopkins, and Nancy Keane. [San Francisco]: Tomcat Press, 1994. 32. [First publication see ref. # 196]

1995

[233] "Post-Card 1995." *Poetry at the 33* 1.1 (Fall 1995): n. p.

1996

[234] "The Case of the Disappearing Couch." *The Typewriter* 5 (Feb. 1996): n.p.
[235] "Bonnie." *The Mendonesian* 2: 9 (March 1996): 14.
[236] "a new view of matter" [Incipit]. *Vegetable Gourmet Cookbook and Wildcrafter's Guide*. Ed. Eleanor & John Lewallen. Mendocino: Sea Vegetable Company, 1996. 100. [First publication see ref. # 184]
[237] "I Always Thought You Black." [excerpt] *Poetry at the 33 Review* 2 (Spring 1996): 28–30.
[238] "Bobby Kaufman." [Excerpt from "I Always Thought You Black"] *Beatitude* 35 (1996): n.p.
[239] "Excerpts" [from "I Always Thought You black", "Full Circle", "Single Out"]. *Contemporary Authors: Autobiography Series* 24 (1996): 325–353.
[240] "Post-Card 1995." *Poetry Now: Sacramento's Literary Calendar & Review* 2.2 (1996): 4. [First publication see ref. # 233]

1997

[241] "Speak for Your Self." *Uvtm* (Jan. 1997): n.p.
[242] "Speak for Your Self." *Sacramento Mufon* 6 (February/March 1997): 5. [First publication see ref. # 241]
[243] "Beast – Be a Saint." *North Beach Now* 11.2 (Feb. 1997): 10. [First publication see ref. # 196]
[244] "Carol." *Sojourn: Passing Spirit, Culture & Knowledge* 2 (Spring 1997): 20.
[245] "For Madeline Gleason." *Sojourn: Passing Spirit, Culture & Knowledge* 2 (Spring 1997): 20. [First publication see ref. # 119]
[246] "My life – a crazy quilt of miracles." *Sojourn: Passing Spirit, Culture & Knowledge* 2 (Spring 1997): 22–23.
[247] "i see you city" [Incipit]. *Awaa-Te* 3 (Spring 1997): 26.
[248] "Seasons." *Awaa-Te* 3 (Spring 1997): 27.
[249] "I Always Thought You Black." [Excerpt] *Bombay Gin: Annual literary magazine of the Jack Kerouac School of Disembodied Poetics* 5.1 (1997): 79–81.
[250] "From The Brink." [Excerpts] *A Different Beat: Writings by Women of the Beat Generation.* Ed. Richard Peabody. London: Serpent's Tail, 1997. 219–224.

1998

[251] "Speak for Your Self." *Poetry at the 33 Review* 3 (1998): 57. [First publication see ref. # 241]
[252] "Mother's Day 1997." *Poetry at the 33 Review* 4 (1998) [special iss. "Commemorative Issue Celebrating 25 Years of Poetry Flash"]: 12.
[253] "Confrontation." *Optimism* 26 (May 1998): 17.
[254] "Beast – Be a Saint." *Optimism* 26 (May 1998): 18. [First publication see ref. # 196]
[255] "Once Upon a Time a Bear. / Es war einmal ein Bär." [Here in German and English, untitled with the incipits "Es war einmal ein Bär" and "There was once upon a bear"] *Optimism* 26 (May 1998): 18.
[256] "Chi Chi Chi." *Optimism* 26 (May 1998): 18.
[257] "One Night." *Gargoyle* 41 (1998): 184.
[258] "Twenty-Third Day." [excerpt] *Speak* (San Francisco) July/Aug. 1998: n.p. [First publication see ref. # 6]
[259] "Beast – Be a Saint." *Speak* (San Francisco) (July/Aug. 1998): n.p. [First publication see ref. # 196]
[260] "Train Song for Jack Micheline (11/6/29–2/27/98)." *The Mendocino Beacon* 20 Aug. 1998: 12.
[261] "I Always Thought You Black." [excerpt] *Discourse: Theoretical Studies in Media & Culture* 20:1–2 (Winter/Spring 1998): 245–250. [Special issue: "The Silent Beat"]
[262] "Beast – Be a Saint." *Kerouac Connection* 29 (1998) [Special Issue for the Stockholm Poetry Festival "Spoken Word – Free the Word"]: 33. [First publication see ref. # 196]
[263] "Listen Papa." *Kerouac Connection* 29 (1998) [Special Issue for the Stockholm Poetry Festival "Spoken Word – Free the Word"]: 34.
[264] "Mother's Day 1997." *Kerouac Connection* 29 (1998) [Special Issue for the Stockholm Poetry Festival "Spoken Word – Free the Word"]: 35. [First publication see ref. # 252]
[265] "A View from Albion." *Kerouac Connection* 29 (1998) [Special Issue for the Stockholm Poetry Festival "Spoken Word – Free the Word"]: 35. [First publication see ref. # 215]
[266] "The Lover's Journey." *X-Ray* 7 (Fall 1998): n.p.

[267] "June 23rd 1977: Anna Akhmatova." *Wood, Water, Air and Fire: The Anthology of Mendocino Women Poets.* Ed. Sharon Doubiago, Devreaux Baker, and Susan Maeder. Comptche: Pot Shard Press, 1998. 300. [First publication see ref. # 7]
[268] "Listen Papa." *Wood, Water, Air and Fire: The Anthology of Mendocino Women Poets.* Ed. Sharon Doubiago, Devreaux Baker, and Susan Maeder. Comptche: Pot Shard Press, 1998. 301. [First publication see ref. # 263]
[269] "Mother's Day 1997." *Wood, Water, Air and Fire: The Anthology of Mendocino Women Poets.* Ed. Sharon Doubiago, Devreaux Baker, and Susan Maeder. Comptche: Pot Shard Press, 1998. 302. [First publication see ref. # 252]
[270] "Wendy." *Wood, Water, Air and Fire: The Anthology of Mendocino Women Poets.* Ed. Sharon Doubiago, Devreaux Baker, and Susan Maeder. Comptche: Pot Shard Press, 1998. 302.
[271] "Alene." *Wood, Water, Air and Fire: The Anthology of Mendocino Women Poets.* Ed. Sharon Doubiago, Devreaux Baker, and Susan Maeder. Comptche: Pot Shard Press, 1998. 302–303.
[272] "Elaine." *Wood, Water, Air and Fire: The Anthology of Mendocino Women Poets.* Ed. Sharon Doubiago, Devreaux Baker, and Susan Maeder. Comptche: Pot Shard Press, 1998, 303.
[273] "Flame." *Wood, Water, Air and Fire: The Anthology of Mendocino Women Poets.* Ed. Sharon Doubiago, Devreaux Baker, and Susan Maeder. Comptche: Pot Shard Press, 1998. 303–304. [First publication see ref. # 231]
[274] "Carol." *Wood, Water, Air and Fire: The Anthology of Mendocino Women Poets.* Ed. Sharon Doubiago, Devreaux Baker, and Susan Maeder. Comptche: Pot Shard Press, 1998. 304. [First publication see ref. # 244]
[275] "Bonnie." *Wood, Water, Air and Fire: The Anthology of Mendocino Women Poets.* Ed. Sharon Doubiago, Devreaux Baker, and Susan Maeder. Comptche: Pot Shard Press, 1998. 305. [First publication see ref. # 204]
[276] "I Always Thought You Black." [Excerpts] *Awaa-Te* 4 (Autumn 1998): 52–53.
[277] "a new view of matter" [Incipit]. *Environmental Alternatives* (1998): 1. [First publication see ref. # 184]

1999

[278] "Requiem for Brew Moore." *5th North Beach Jazz Festival: July 25th – August 1st 99* [program folder]: 8. [First publication see ref. # 84]
[279] "Meeting Jack Micheline – Head On." *Ragged Lion: A Tribute to Jack Micheline.* Ed. John Bennet. NYC: The Smith Publishers / Ellensburg: Vagabond Press, 1999. 24–25.
[280] "Jack Micheline: 67 Poems for Downtrodden Saints." [Blurb] *Beat Scene* 33 (1999): n.p.
[281] "1985." Optimism 29 (May/June 1999): n.p.
[282] "For Madeline Gleason." *The Outlaw Bible of American Poetry.* Ed. Alan Kaufman. New York: Basic Books, 1999. 534. [First publication see ref. # 119]

2000

[283] "Train-Song for Jack Micheline." *Eavesdropping on the Muse: Mission & North Beach Poets.* Ed. Rob Brooker. San Francisco: Luna's Press, 2000. N.p. [First publication see ref. # 260]
[284] "One More Step West Is the Sea." [Excerpt] *The San Francisco Bay Guardian* 34.41 (2000): n.p. [First publication see ref. # 1]

2001

[285] "there – there" [Incipit]. *The Café Review* 12 (Fall 2001): 17.
[286] "is this really dark" [Incipit]. *The Café Review* 12 (Fall 2001): 27.

[287] "For Woody the Painter Elwood Miller." *The Café Review* 12 (Fall 2001): 40.

2002
[288] "Hear Us". *Captain Fathom's Fables: Interview Sketches Fables Notes Poetry.* By Alan Graham. Mendocino: Pacific Transcriptions, 2002. 57. [First publication see ref. # 204]
[289] "Light." *And the Beat Goes On: A Poetry Anthology.* Ed. Bryn Fortey. Newport: Forty Winks Press, 2002. 25. [First publication under the title "Two B," here without the first three lines, see ref. # 5.]
[290] "Mother's Day 1997." *And the Beat Goes On: A Poetry Anthology.* Ed. Bryn Fortey. Newport: Forty Winks Press, 2002. 26. [First publication see ref. # 252]
[291] "Train Song for Jack Micheline." *And the Beat Goes On: A Poetry Anthology.* Ed. Bryn Fortey. Newport: Forty Winks Press, 2002. 27. [First publication see ref. # 260]
[292] "Ten Ten." *The Café Review* 13 (Spring 2002): 20–22.
[293] "I Hear with Love." *Jazzsteps* 1.10 (2002): 8. [First publication see ref. # 10]
[294] "Confrontation." *Outlaw Magazine* 1 (Winter 2002): 5. [First publication see ref. # 253]
[295] "Beast – Be a Saint." *Outlaw Magazine* 1 (Winter 2002): 11. [First publication see ref. # 196]
[296] "Soap & Silver Clean Like Rain." *Outlaw Magazine* 1 (Winter 2002): 27.
[297] "Speak for Yourself." *Outlaw Magazine* 1 (Winter 2002): 32. [First publication see ref. # 241.]

2003
[298] "Her Number Was Not Called." *A Letter to the Stars: Briefe in den Himmel: Schüler schreiben Geschichte.* Ed. Alfred Worm, Harald Krassnitzer, Andreas Kuba, Manfred Lang, Regine Muskens, and Josef Neumayr. Vienna: Verein Lernen aus der Zeitgeschichte, 2003. N.p.

2004
[299] "I Always Thought You Black." [Excerpt] *Outlaw Magazine* 6 (Spring 2004): 26–27.
[300] "j'ai toujours pensé que tu étais noir(e)." [Translation of "I Always Thought You Black", excerpt] *Le Journal Des Poètes* 73.2 (2004): 5.

2005
[301] "i hear the voice of my closest friend" [Incipit, extract of a letter to the editors] [a.k.a. "Susi Oh Susi"]. *Einblicke. Ausblicke: 10 Jahre Nationalfonds.* Ed. Renate S. Meissner. Vienna: Verlag Jugend und Volk, 2005. 165.

2006
[302] "Tidal-Poem." *Poetry & Jazz: 4th Annual Redwood Coast Whale & Jazz Festival* (Point Arena) [program folder] (2006): 4.
[303] "Get A Life." *Transit: A Little Magazine of the Beat Generation* 17 (2006): n.p.
[304] "Chi Chi Chi." *Transit: A Little Magazine of the Beat Generation* 17 (2006): n.p. [First publication see ref. # 256]
[305] "Ten Ten." *Poets, Politics, and Divas: Ira Nowinski's San Francisco.* Preface by Jack Euw. Foreword by Rebecca Solnit. Berkeley: The Bancroft Library, 2006. N.p. [First publication see ref. # 292]

2007

[306] "Beast – Be a Saint." *Beat: Christopher Felver: Photographs/Commentary*. San Francisco: Last Gasp [2007]. N.p. [First publication see ref. # 196]
[307] "A Ballad for Our Time." *The 16th & Mission Review* 4 (Nov./Dec 2007), 21–22. [First publication see ref. # 205]
[308] "I Am Calling You, Laura Ulewicz (1930–2007)." *Gargoyle* 54 (2007): 137–138.
[309] "For Carol Bergé (1928–2006)." *Gargoyle* 54 (2007): 139.
[310] "For Philip Lamantia." *Gargoyle* 54 (2007): 140.

2008

[311] [Quotes from: "For Bobby Kaufman", "Something Current", "Post-Card 1995", "The Brink", "Single Out" (2)] *The Poetry Oracle*. By Amber Guetebier and Brenda Knight. San Francisco: Consortium of Collective Consciousness, 2008. 35+.

2009

[312] "For Sandy Goulart: 5/14/1940–1/7/2009." A Grand & Glorious Music Extravaganza: A Gift from Sandra Goulart: May 14, 1940 – January 7, 2009: And a Memorial Celebration of Her Life. Albuquerque, NM: 2009. N.p.
[313] "Al Robles 2/16/1930–5/2/2009." Transit: A Little Magazine of the Beat Generation 22 (Autumn 2009): n.p.
[314] "For Philip Lamantia." Transit: A Little Magazine of the Beat Generation 22 (Autumn 2009): n.p. [First publication see ref. # 310]
[315] "Spring Begins." Transit: A Little Magazine of the Beat Generation 22 (Autumn 2009): n.p.
[316] "Poem." ["the animals have come down" Incipit]. Beatitude: Golden Anniversary 1959–2009. Ed. Latif Harris, Co-Editor Neeli Cherkovski, Associate Editors John Landry and Michael Rothenberg. San Francisco: Latif Harris, 2009. N.p. [First publication see ref. # 27]

2010

[317] "M & M." *The Kenning Anthology of Poets Theater 1945–1985*. Ed. Kevin Killian and David Brazil. Chicago: Kenning Editions, 2010. 221–231.
[318] "For Carol Berge." *Transit: A Little Magazine of the Beat Generation* 23 (Spring 2010): n.p. [First publication see ref. # 309]
[319] "I Am Calling You: Laura Ulewicz (1930–2007) Into Orbit." *Transit: A Little Magazine of the Beat Generation* 23 (Spring 2010): n.p. [First publication see ref. # 308.]
[320] "Sutter Marin Cherished Friend." *Transit: A Little Magazine of the Beat Generation* 23 (Spring 2010): n.p.
[321] "Meeting Jack Micheline – Head On." *Transit: A Little Magazine of the Beat Generation* 23 (Spring 2010): n.p. [First publication see ref. # 279]
[322] "2010." *Sparring with Beatnik Ghosts* 4 (2010): n.p.
[323] "Beast – Be a Saint." *Sparring with Beatnik Ghosts* 4 (2010): n.p. [First publication see ref. # 196.]
[324] "ruth weiss." *Sparring with Beatnik Ghosts* 4 (2010): n.p. [autobiographical poem, not identical with the poem of the same name in *For These Women of the Beat Generation*]
[325] "Bête – Soit Un Saint." [French translation of "Beast – Be a Saint"] *Action Poétique* 200 (2010): 22. [First publication see ref. # 196]
[326] "La Fête Des Mères 1997." [French translation of "Mother's Day 1997"] *Action Poétique* 200 (2010): 22. [First publication see ref. # 252]

[327] "Chi Chi Chi." [French translation of "Chi Chi Chi"] *Action Poétique* 200 (2010): 23. [First publication see ref. # 256]
[328] "Chant De Train." [French translation of "Train Song for Jack Micheline"] *Action Poétique* 200 (2010): 23. [First publication see ref. # 260]
[329] "C'Est Vraiment Vrai." [French translation of "This Is Really Real"] *Action Poétique* 200 (2010): 24. [First publication see ref. # 209]
[330] "2009." [French translation of "2009"] *Action Poétique* 200 (2010): 24.
[331] "De Moi A Toi." [French translation of "From Me to You"] *Action Poétique* 200 (2010): 25.
[332] "1967." [French translation of "1967"] *Action Poétique* 200 (2010): 26.

2011
[333] "I Am Calling You Laura Ulewicz (1930–2007): Obit Into Orbit." *Sparring with Beatnik Ghosts* 2.1 (2011): 143. [First publication see ref. # 308]
[334] "For Philip Lamantia." *Sparring with Beatnik Ghosts* 2.1 (2011): 144. [First publication see ref. # 310]
[335] "For Carol Bergé." *Sparring with Beatnik Ghosts* 2.1 (2011): 144. [First publication see ref. # 309]
[336] "2011." *Sparring with Beatnik Ghosts* 2.1 (2011): 145.
[337] "Eleventh Day." *Askew* 11 (Fall/Winter 2011/2012): n.p. [First publication see ref. # 6]

2012
[338] "Dia Ocho" [Spanish translation of "Eighth Day"], transl. by José Vicente Anaya. *Traslaciones: Poetas traductores 1939–1959*. Ed. Tedi López Mills. Mexico City: Fondo de Cultura Económico, 2012. 159–165. [First publication see ref. # 6]
[339] "2012." *Sparring with Beatnik Ghosts* 2.2 (2012): n.p.
[340] "ruth weiss." *Sparring with Beatnik Ghosts* 2.2 (2012): n.p.

2014
[341] "Jane Bowles." [translation into Swedish] *Lyrik vännen* 61.1–2 (2014): 28. [First publication see ref. # 9]
[342] "Ilse Klapper." [translation into Swedish] *Lyrik vännen* 61.1–2 (2014): 28. [First publication see ref. # 9]
[343] "Joan Vollmer Adams Burroughs." [translation into Swedish] *Lyrik vännen* 61.1–2 (2014): 28. [First publication see ref. # 9]
[344] "Elise Cowen." [translation into Swedish] *Lyrik vännen* 61.1–2 (2014): 29. [First publication see ref. # 9]
[345] "Janine Pommy Vega." [translation into Swedish] *Lyrik vännen* 61.1–2 (2014): 29. [First publication see ref. # 9]
[346] "Jan Kerouac." [translation into Swedish] *Lyrik vännen* 61.1–2 (2014): 29. [First publication see ref. # 9]

2015
[347] "At 85 in 2013." *Milk: A Poetry Magazine* 3–4 (Spring 2015): n.p. [First publication see ref. # 19]
[348] "Forty-First Day: The Return." *Milk: A Poetry Magazine* 3–4 (Spring 2015): n.p. [First publication see ref. # 19]

[349] "For Paul Blake: May 16, 1945 – October 2, 2014." *Gazette: A Magazine of Stories, Poems and Personal Narratives* (July–September 2015): 15.
[350] "Antonia Lamb: 10/03/43 – 9/9/13." *Gazette: A Magazine of Stories, Poems and Personal Narratives* July–September 2015: 15.
[351] "Second Day" / "Segundo Día." [original and spanish translation] *Beat Attitude: Antología de mujeres poetas de la generación beat*. Ed., transl., and prefaced by Annalisa Marí Pegrum. Madrid: Bartleby Editores, 2015. 118 – 123. [First publication see ref. # 6]
[352] "Eighth Day" / "Octavo Día" [original and spanish translation] *Beat Attitude: Antología de mujeres poetas de la generación beat*. Ed., transl., and prefaced by Annalisa Marí Pegrum. Madrid: Bartleby Editores, 2015. 124 – 131. [First publication see ref. # 6]
[353] "Post-Card 1995" / "Postal 1995" [original and spanish translation] *Beat Attitude: Antología de mujeres poetas de la generación beat*. Ed., transl., and prefaced by Annalisa Marí Pegrum. Madrid: Bartleby Editores, 2015. 132 – 137. [First publication see ref. # 233]
[354] "Ten Ten" / "Diez Diez" [original and spanish translation] *Beat Attitude: Antología de mujeres poetas de la generación beat*. Ed., transl., and prefaced by Annalisa Marí Pegrum. Madrid: Bartleby Editores, 2015. 138 – 145. [First publication see ref. # 292]

2018

[355] "Second Day" / "deuxième jour" [original and french translation] *Beat Attitude: Femmes Poètes de la Beat Generation*. Ed. Annalisa Marí Pegrum, Sébastien Gavignet. Paris: Éditions Bruno Doucey, 2018. 108 – 113. [First publication see ref. # 6]
[356] "Eighth Day" / "huitième jour" [original and french translation] *Beat Attitude: Femmes Poètes de la Beat Generation*. Ed. Annalisa Marí Pegrum, Sébastien Gavignet. Paris: Éditions Bruno Doucey, 2018. 114 – 121. [First publication see ref. # 6]
[357] "Post-Card 1995" / "carte postale 1995" [original and french translation] *Beat Attitude: Femmes Poètes de la Beat Generation*. Ed. Annalisa Marí Pegrum, Sébastien Gavignet. Paris: Éditions Bruno Doucey, 2018. 123 – 127. [First publication see ref. # 233]
[358] "Ten Ten" / "dix dix" [original and french translation] *Beat Attitude: Femmes Poètes de la Beat Generation*. Ed. Annalisa Marí Pegrum, Sébastien Gavignet. Paris: Éditions Bruno Doucey. 2018. 128 – 135. [First publication see ref. # 292]
[359] "Phyllis Holliday: 6/20/1935 – 9/16/2017." *Haight Ashbury Literary Journal* 34 (Summer 2018): 2.
[360] "In Hall in Tirol." [german translation] *Austrian Beat*. Ed. Elias Schneitter and Helmuth Schönauer. Zirl: Edition Baes, 2018. 56.
[361] "Interview in Österreich." [german translation] *Austrian Beat*. Ed. Elias Schneitter and Helmuth Schönauer. Zirl: Edition Baes, 2018. 57.
[362] "Die Begegnung mit Jack Micheline—Direkt." [german translation of "Meeting Jack Micheline – Head On"] *Austrian Beat*. Ed. Elias Schneitter and Helmuth Schönauer. Zirl: Edition Baes, 2018. 58 – 60. [First publication see ref. # 279]
[363] "For These Women of the Beat." [excerpts, german translation] *Austrian Beat*. Ed. Elias Schneitter and Helmuth Schönauer. Zirl: Edition Baes, 2018. 61 – 64. [First publication see ref. # 9]
[364] "As the Wheel Turns" *Austrian Beat*. Ed. Elias Schneitter and Helmuth Schönauer. Zirl: Edition Baes, 2018. 65.
[365] "2018: The Circus Is in Town" *Austrian Beat*. Ed. Elias Schneitter and Helmuth Schönauer. Zirl: Edition Baes, 2018. N.p. [printed on back cover]

2020

[366] "re: Election. Fact or Fiction." *Triëdere* (Vienna) 22 (2020): 129.

I.3 Audio releases

[367] *Poetry & All That Jazz*, Vol 1 & 2. Audio Cassette, USA: Awarehouse, 1994.

[Content: "[Excerpts] from The Brink"; "This Is Really Real"; "Confrontation"; "13 Haiku"; "Desert Journal, 26th Day"; "Desert Journal, 19th Day"; "Desert Journal, 14th Day"; "Billy Cannon"; "Turnabout"]

[368] *A New View of Matter*. Audio-CD, USA: Awarehouse, 2000.

[Content: "A New View of Matter"; "Beast – Be a Saint"; "This Is Really Real"; "Mother's Day 1997"; "Listen Papa"; "21st Day"; "15th Day"; "Train Song for Jack Micheline"; "Chi Chi Chi"; "Light" [excerpt]; "Africa" [excerpt]; "Atone at One"]

[369] *Poetry & All That Jazz Vol. 3*. Audio CD, USA: Awarehouse 2004.

[Content: "Opening My Third Ear"; "For These Women of the Beat"; "Bobby Kaufman"; "Post-Card 1995"]

[370] *3 Farben: Weiss – Music of Peace*. Audio CD. Vienna, Austria: ORF 2004.

[Content: "White Is All Colors"; "Light Works" by ruth weiss, and compositions by Mayako Kubo, Julia Hülsmann, Ming Wang]

[371] *Nachklang – Widerhall*. Audio CD. Leonding, Austria: Kult-Ex 2007.

[Content: "Her Number Was Not Called" and 38 tracks by other artists]

[372] *Turnabout*. Audio CD, Albion, CA, 2010. [self-published]

[Content: "Turnabout."]

[373] *Make Waves: Jazz & Poetry*. Audio CD. Vienna: Edition Exil, 2013.

[Content: "2013"; "Once Upon a Time a Bear"; "Get a Life"; "Beast – Be a Saint"; "Fire Works"; "From Me to You"; "1967"; "Chi Chi Chi"; "I Hear with Love"; "Speak for Your Self"; "Fourteenth Day"; "Twenty-Sixth Day"; "Twenty-Ninth Day"; "Thirty-Ninth Day"]

[374] Jack Micheline: *Give America a Break*. Double Vinyl LP, San Francisco: Unrequited Records, 2014.

[Content: "Train Song for Jack Micheline" along with 32 tributes by other poets on Side C and D of the release and readings by Jack Micheline on sides A and B]

[375] *Jazz & Haiku – ruth weiss Live in Oakland*. [Recording from 2017.] Audio CD. Vienna: Literaturhaus Wien / Absurdia Records, 2018.

[Content: "Walking a Tightrope & Other Haiku"; "Writing Haiku with Jack Kerouac"; "From 'Chopsticks'"; "San Francisco 1957"; "13 Senryu"; "A Fool's Journey"; "For These Women of the Beat"; "Banzai"; "41st Day"; "Speak for Yourself"]

[376] *Live at the Beat Museum 2018 & Other Recordings from the Archive*. Audio Cassette. The Hague, Netherlands: Counter Culture Chronicles, 2019.

[Content: Side A: ruth weiss Live at the Beat Museum, July 28, 2018: "2018"; "90 in 2018"; "Bypass Linz"; "For These Women of the Beat Generation [Excerpts]"; "Ten Ten"; "18th Day"; "[Excerpts from] I Always Thought You Black" / Side B: ruth weiss & Bruce Lippincott, Santa Fe, November 4, 1969: "39th Day"; "36th Day"; "25th Day"; ruth weiss at Radio KPOO, October 27, 1973: "Scott Runyon"; "Interview about Surprise Voyage"; "Single Out"; "Interview"]

I.IV Film works and published performances on film

[377] *The Brink*. 16 mm black & white, 40 min. 1961. (Location of the original film: Pacific Film Archive, UC Berkeley)

[378] *Poetry & All That Jazz,* Vol. 1, Sonoma Salute to the Artist of Calif., 1990, VHS.

[379] *Poetry & All That Jazz,* Vol. 2, West Coast Jazz Festival, Kimball East, Emeryville, CA, 1993, VHS.

[380] *Women of the Beat Generation*, San Bruno, CA, 1996, VHS.

[381] *Las Cuevas de Albion*. Dir. ruth weiss and Alfonso Gordillo, 12 min., 2001, VHS.

[382] *Ibéria*. Dir. Eddy Falconer, 65 min., 2006, DVD. [features ruth weiss acting for three minutes to a 1990 recording of "19th Day"]

[383] *ruth weiss Meets Her Prometheus*. Dir. Frederick Baker, 18 min., 2007, DVD.

[384] *Her Number Was Not Called / ruth weiss + Hal Davis with Prometheus*, 2013. Dir. Andreas Holleschek, 3 min., Vienna: Edition Exil, DVD.

[385] *ruth weiss at Monroe, San Francisco* (DVD, 2016). Dir. unknown; produced by the Beat Museum, 131 minutes.

II List of works in alphabetical order

10 [8]
104 [129]
1112 [146]
1127 [8]
11684 [8]
12 19 84 [136]
120 [151]
1223 [132; 154]
13 Haiku [8, 367]
13 Senryu [375]
1327 [8]
14 [8]
1416 – 9 [149; 169]
16 – 17 [134]
18 [156]
1967 [16; 332; 373]
1984 [131]
1985 [281]
1988 [182]
1991 [192, 193]
20 [123; 145]
2009 [17; 330]
2010 [17; 322]
2011 [17; 336]
2012 [17; 339]
2013 [19; 373]
2018: The Circus Is In Town [376]
21485 [8]
21684 [150]
21884 [155; 168]
23 [126]
24 [124; 130]
31 [135; 139]
3584 [157; 166]
4 [8; 144; 167]
41 Dragon Steps [1]
414 [153]
41684 [172]
4583 [8]
47 [158]
484 [147]
5 [170]
51083 [141]
5x7 – 5x12 [17; 185]

62088 [183]
6283 [8]
63 [194]
713 [8]
8186113 [8]
8383 [8]
84 [8; 152]
90 in 2018 [376]
985 [164]
A la Blake [2]
Africa [11; 42; 117; 368]
Africa [11]
Afterfact [190]
Against All the Odds [202]
"air air air air air" [Incipit] [21]
Al Robles [313]
Alene [271]
All [5]
Altar-Piece [16; 121; 229]
Ana in Mexico [2]
Ana in the English Lesson [2]
Anaïs [7]
Ancestor Figure [3]
And What About Money? [97]
"the animals have come down" [Incipit] [27; 41; 316]
Anne Waldman [9]
Antonia Lamb: 10/03/43 – 9/9/13 [350]
Any [2]
"apple of god's eye" [Incipit] [173]
As the Wheel Turns [364]
At 85 in 2013 [19, 347]
Atone at One [180; 368]
Author's Note [9]
"autumn leaves falling" [Incipit] [201]
Aya Tarlow [9]
"the baby is always white" [Incipit] [13]
"the baby is born" [Incipit] [13]
A Balancing Act [142]
A Ballad for Our Time [205; 210; 307]
Banzai [21; 375]
Barbara Guest [9]
Beast – Be a Saint [10; 16; 196; 232; 243; 254; 259; 262; 295; 306; 323; 325; 368; 373]
"the bells are ringing" [Incipit] [13]
Belt [3]
Between [105]
Beverly [2]
Beyond the Palace Walls [16]

Big Sur [28]
Billie [107]
Billie Holiday [9; 17; 363]
Billy Cannon [189; 221; 367]
The Blind Voyage [77]
Blue In Green [4]
The Blues Singer [46]
Bomkauf You Did It Again [219; 226]
Bonnie [235; 275]
Boo [2]
Bowl with Carved Figures As Handles [3]
Brenda Frazer [9; 17]
Brenda Knight [9; 17]
The Brink [poem] [7; 10; 74; 250; 367]
The Brink [film] [377]
The Brink [Film, 16 mm b&w, 40 min.] (1961)
"A broken stairway suspended" [Incipit] [25]
"butterfly on blossom" [Incipit] [13]
Bypass Linz [17; 376]
Can't Stop the Beat: The Life and Words of a Beat Poet [15]
Candle & Leaves [5; 90]
Carol [244; 274]
Carolyn Cassady [9; 17]
Carving Use In Mortuary Ceremonies [3]
The Case of the Disappearing Couch [234]
"a cave in the dark continent" [Incipit] [13]
Centered [5; 94]
"cerberus" [Incipit] [38]
Chi Chi Chi [10; 16; 256; 304; 327; 368; 373]
The Children – Yes! [16]
Chopsticks [7; 375]
Circle V Motel, Independence, California [116]
"the claw of the beast" [Incipit] [1998]
Cloud Calling Sisters [203]
Cockie with Earth-Paintings [2]
Compass [7; 15; 40]
Competition [159]
Confrontation [10; 253; 294; 367]
Coyote [5; 93; 138; 224]
"crag" [Incipit] [4]
"crane dance step by step" [Incipit] [127]
Current Events or A Political Poem or I Hope This Is Not True Tomorrow [83]
Cynthia [2[
Dagger Shaped As Bird [3]
"the dance in the flame" [Incipit] [21]
Dean Antony [5; 112]
Decorated Bowl Shaped As Bird [3]

Decorated Comb [3]
Dedication [5]
Dedication [10] [not identical to above]
Deirdre [5]
"delicate bird" [Incipit] [102]
Denise Levertov [9; 17]
Desert Journal [6]
Diane [2]
"diane" [Incipit] [4]
Diane di Prima [9; 17]
District God [3]
Don Murray [220]
Dora [2]
Drum Base Figure [3]
Edie Parker Kerouac [9; 17]
Eighteenth Day [6; 19; 376]
Eighth Day [6; 19; 163; 338; 352; 356]
Eileen Kaufman [9; 17]
Einen Schritt weiter im Westen ist die See [18]
Elaine [272]
Eleventh Day [6; 18; 86; 337]
Elise Cowen [9; 17; 344]
"enough words! start brush" [Incipit] [21]
Epilogue: Message from Ben [162]
Elsie John [9; 17]
Eva [2]
Every Suicide is Fratricide: Nine Fragments for Kitty [72]
Exodus -- 1136 [30]
"ezra talked to me last night" [Incipit] [78]
Fairy Tale [60]
Fifteenth Day [6; 19; 368]
Fifth Day [6; 19]
Fifty-One [120]
Figs [14]
Fire Works [16; 373]
First Day [6; 19]
Five A [5]
Five B [5]
Five C [5]
Flame [231; 273]
Flamenco for Harriet and Larry [207]
Flower in Rock [47]
Flugschrift [21]
A Fool's Journey [16; 21; 375]
A Fool's Journey / Die Reise des Narren [16]
For Benfaral Matthews [115; 161]
For Bobby Kaufman [186; 369]

For Carol Bergé [18; 309; 318; 335]
"For Health of Ma Earth" [Incipit] [195]
For Jimmie Lowe Last Night at That Pub in San Francisco [98]
For Madeline Gleason [119; 245; 282]
For Otter and Sutter [122]
For Paul Blake: May 16, 1945 – October 2, 2014 [349]
For Philip Lamantia [17; 310; 314; 334]
For Sandy Goulart [312[
For T. R. One Hot Noon on Reading [53]
For These Women of the Beat [9; 17; 363; 369; 375; 376]
For Woody The Painter Elwood Miller [287]
Foreword [9]
Foreword [11] [not identical with above]
Fortieth Day [6; 18; 133]
Forty-First Day – The Return [19; 348; 375]
Four [7]
Four A [5]
Four B [5]
Four for Sutter [177]
Fourteenth Day [6; 19; 367; 373]
Fourth Day [6; 19]
From Me to You [16; 331; 373]
Fuck Me or Graffiti [79]
Full Circle [12; 239]
Full Circle / Ein Kreis vollendet sich [12]
Gabrielle 'Mèmère' Kerouac [9; 17]
Gallery of Women [2]
The Garden Within [59]
"the gate is open" [Incipit] [13]
George Abend [146]
Get a Life [16; 303; 373]
Ghia [5; 91]
"ghost-scents" [Incipit] [4]
Giulletta [2]
"green green green the hill" [Incipit] [21]
"green thoughts" [Incipit] [4]
Guardian Angel [58]
Gui de Angulo [9; 17]
"Hear Us" [204; 211; 288]
Helen [2]
Helen Adam [9; 17]
Helen Hinkle [9; 17]
Her Number Was Not Called [14; 298; 371; 384]
"Her presence assailed the room" [Incipit] [23]
"here harry" [Incipit] [73]
Hettie Jones [9; 17]
Horizon [81]

I Always Thought You Black [10; 15; 238; 239; 249; 261; 276; 300; 376]
I Am Calling You – Laura Ulewicz, Obit Into Orbit [17; 308; 319; 333]
"i hear the voice of my closest friend" [Incipit] [301]
I Hear with Love [10; 16; 293; 373]
"i see you city" [Incipit] [247]
Ibéria [382]
Idell [2]
"if i were the rooster" [Incipit] [222]
"if the spiderweb" [Incipit] [70]
Ilse Klapper [9; 17; 342]
"in a breezy wink" [Incipit] [4]
In A Japanese Tree Garden [3]
In Hall in Tirol [17; 360]
"in search for her word" [Incipit] [21]
Incised Club [3]
Infinite [140; 171]
"internal weather" [Incipit] [125]
Interview in Austria [17]
An Invitation [16]
Irene [5]
"is this really dark" [Incipit] [21; 286]
"it is told" [Incipit] [13]
"It's a rape-night" [Incipit] [39]
Iva [2]
J. [114]
Jack Micheline: 67 Poems for Downtrodden Saints [280]
Jan Kerouac [9; 17; 346]
Jane Bowles [9; 17; 341]
Janine Pommy Vega [9; 17; 345]
Jay DeFeo [9; 17]
Jean [2]
Joan [2]
Joan Brown [9; 17]
Joan Haverty Kerouac [9; 17]
Joan Vollmer Adams Burroughs [9; 17; 343]
Joanna McClure [9; 17]
Joanne Kyger [9; 17]
John Hoffman [4]
Josephine Miles [9; 17]
Joyce Johnson [9; 17]
Judy [2]
June 23rd, 1977 – Anna Akhmatova [7; 10; 267]
Kaisik [5]
Karen Blixen [48]
Kitty [2]
L'Avventura [55]
La Dolce Vita [55]

Las Cuevas de Albion [381]
Laura [2]
The Legacy of Prometheus [14]
Lemonspiel [4]
Lenore Kandel [9; 17]
Leslie [2]
Light [5; 10; 18; 20; 289; 368]
Light and Other Poems [5; 20]
The Light Rose Pink [174]
Light Works [16; 370]
Lime-Tube Stopper [3]
Listen Papa [10; 16; 263; 268; 368]
Litany for Wanda [213]
"Loose-flapped in cloaked dignity" [Incipit] [24]
The Lover's Journey [266]
Luanne Henderson & Anne Murphy [9; 17]
M & M [10; 14; 317]
"Ma Earth stop quaking" [Incipit] [230]
Madeline Gleason [9; 17]
Make Waves [373]
"making room for love" [Incipit] [21]
Maple-Bridge Night-Mooring [a.k.a. Maple Bridge] [32; 52]
Marianne [2]
The Marvelous Mind That I Am [2]
Mary Fabilli [9; 17]
Mary Norbert Körte [9; 17]
Meeting Jack Micheline – Head On [17; 279; 321; 362]
Miracle Burn [87]
"The mists swam across the night" [Incipit] [26]
Modelled Mask [3]
Moss [5]
Mother's Day 1997 [10; 16; 252; 264; 269; 290; 326; 368]
'Msusumu' Canoe Prow Figue [3]
Mutti [5]
My Life – A Crazy Quilt of Miracles [246]
Natalie Jackson [9; 17]
"a new view of matter" [Incipit] [10; 184; 236; 277; 368]
A New View of Matter / Nový pohled na věc [10]
Nineteenth Day [6; 19; 367]
Ninth Day [6; 19]
No Dancing Aloud [14; 104]
No Dancing Aloud / Lautes Tanzen nicht erlaubt [14]
No Harvesting Here [a.k.a. "49"] (n.d.)
"Now, we're talking about the end of 1971" [Incipit] [197]
"ocean-marriage" [Incipit] [4; 34]
Oh Ana Ana Oh [2]
Once Upon a Time a Bear [16; 255; 373]

One [5]
"one could say" [Incipit] [71]
"one foot off the cliff" [Incipit] [21]
One Knight and One Day [14]
One More Step West Is the Sea [1; 18; 284]
One Night [257]
Opening My Third Ear [369]
Orion: Toxic Waste [212]
Out with without a Home [225]
Overheard: Coffeshop Talk [43]
Oyster Stew [5; 101]
"Pain Ting" [Incipit] [37]
Painted Bark Cloth [3]
Painted Decorative Carving [3]
"the paper is blank" [Incipit] [13]
A Parallel Planet of People and Places: Stories and Poems [17]
Pattie [2]
Paul [5]
Paul Zipp [217; 227]
"peaks of snow" [Incipit] [13]
Phyllis [2]
Phyllis Holliday: 6/20/1935 – 9/16/2017 [359]
Pier [2]
Pig-Killer Adze-Shaped [3]
Poetry and All That Jazz 1 & 2 [367]
Poetry and All That Jazz 3 [369; 378]
"the point become line" [Incipit] [13]
Post-Card 1995 [15; 233; 240; 353; 357; 369]
"a promise kept" [Incipit] [13]
Rashomon [55; 57]
Ray [85; 128]
re: Election. Fact or Fiction [366]
"red-golden is the crown" [Incipit] [223]
Requiem for Brew Moore [84; 278]
The Return [82]
Rita [2]
The Rock on the Road [5]
"rock the many" [Incipit] [4]
"the rose chose the poet" [Incipit, a.k.a. "Rainer Maria Rilke"] [21]
ruth weiss [9; 17]
ruth weiss [324; 340] [not identical with above]
ruth weiss at Monroe [385]
ruth weiss Speaks to the Editor [12]
"salt-blue" [Incipit] [4]
San Francisco 1957 [375]
Sandy [2]
"say everything you mean" [Incipit] [100]

SB 883 [148]
"the score of the trees" [Incipit] [178; 181]
Scott Runyon [376]
"Scraps & vertical & horizontal" [Incipit] [62]
"THE SEA BRINGS MESSAGE" [Incipit] [21]
Seasons [248]
Seat Shaped as Bird [3]
Second Day [6; 19; 88; 351; 355]
Sena [5; 111; 137]
Seven A [5]
Seven B [5]
"seven stones" [Incipit] [4]
Seven Times Yelapa [29]
Seventeenth Day [6; 10; 18]
Seventh Day [6; 18]
Sharp Up [5]
"she knows what she knows" [Incipit] [13]
Sheila [2]
Shibu Wabi Sabi Furu [65]
"ships of the desert" [Incipit] [13]
Shirley [2]
Single Out [7; 10; 12; 75; 239; 376]
Single Out [7]
Six A [5]
Six B [5]
Sixteenth Day [6; 18; 88]
Sixth Day [6; 19]
'Skakabul' Mortuary Dance Object [3]
The Snake Sez Yesssss / Die Schlange sagt jetzzzzzt [19]
"snowflakes" [Incipit] [13]
Soap and Silver Clean Like Rain [296]
"someone sensitive" [Incipit] [63]
Something Current [7; 18; 31; 118]
Sor Juana [2]
"sound moves solids" [Incipit] [27]
South Pacific [3]
Speak for Your Self [a.k.a. "Speak for Yourself"] [10; 16; 241; 242; 251; 297; 373; 375]
Spirit-Talk [108]
Spring Begins [315]
Stella Sampas [9; 17]
Steps [1]
Still Dora [2]
"the stories are the milky way" [Incipit] [13]
Sueko [2]
Suicide Dreams [16]
Sun Burnt to Kites [49]
Sungtim [66; 188]

Suspension Hook [3]
Suspension Hook With Bird Form [3]
Sutter Marin [17]
Sutter Marin, Cherished Friend [320]
Sylvia Coddington [179]
Tale About Dori or A Lioness With Mane [2]
'Tale' Carved Door Jamb [3]
Ten Ten [15; 292; 305; 354; 358; 376]
Tenth Day [6; 19; 103]
Terry O'Flaherty (Crashed 2/7/87) [175; 176]
"there – there" [Incipit] [21]
"They sip their beer" [Incipt] [22]
Third Day [6; 19]
Thirteenth Day [6; 19]
The Thirteenth Witch [14]
Thirtieth Day [6; 19]
Thirty-Eighth Day [6; 19]
Thirty-Fifth Day [6; 18]
Thirty-First Day [6; 18]
Thirty-Fourth Day [6; 19]
Thirty-Ninth Day [6; 19; 373; 376]
Thirty-Second Day [6; 19]
Thirty-Seventh Day [6; 19]
Thirty-Sixth Day [6; 19; 376]
Thirty-Third Day [6; 19]
This Is Really Real [10; 209; 329; 367; 368]
"those hives of the past" [Incipit] [21]
Three [5]
Tidal-Poem [302]
"Ting A Ling" [Incipit] [35]
To a Poem Called Mary Stagliano [99]
To Dee or Robin's Nest [2]
To Each Other [50]
To Gogo [2]
To Lois and Her Cricket [2]
To Nancy Nefertete [2]
To Phyllis One January [2]
To Sue [2]
To V W [2]
The Tooth of the Tiger [33]
Torch-Song for Prometheus [12]
Train Song for Jack Micheline [10; 17; 260; 283; 291; 328; 368; 374]
Tse-Wah-Te-Ay [5; 113]
Turnabout [10; 18; 200; 367; 372]
Turtle Shaped Platter [3]
Twelfth Day [6; 18]
Twentieth Day [6; 19]

Twenty-Eighth Day [6; 19]
Twenty-Fifth Day [6; 19; 376]
Twenty-First Day [6; 18; 368]
Twenty-Fourth Day [6; 19]
Twenty-Ninth Day [6; 19; 373]
Twenty-Second Day [6; 18]
Twenty-Seventh Day [6; 19]
Twenty-Sixth Day [6; 19; 367; 373]
Twenty-Third Day [6; 10; 18; 258]
Two A [5]
Two B [5]
Two C [5]
Untitled [54]
An Untitled Story [110]
Us [5; 92; 96; 199]
Vickie Russell [9; 17]
A View from Albion [215; 216; 218; 228; 265]
Vita-Sheet [76]
Voicing [160]
Vortex [5]
"Walking a Tightrope" [143]
Walking a Tightrope and Other Haiku [375]
"warm swarms the wind" [Incipit] [4]
Water and Fire [69]
"we burn from the moment born" [Incipit] [64]
Wendy [270]
The White Dove (After a Film by Frantisek Vlacil) [55; 56]
"white dove on the wing" [Incipit] [13]
White Is All Colors [13; 370]
"white light & soft" [Incipit] [13]
Who Will Throw the Pie in Eye? [45]
Who? Me? [68]
"who's going to read your books?" [Incipit] [95]
Why Did You Go to Las Vegas to Die Ron Towe [208]
"Why I write poetry" [Incipit] [214]
"will sing & soar" [Incipit] [61]
Women of the Beat Generation [380]
Words No Words for Dance Fumi [2]
Writing Haiku with Jack Kerouac [375]
"Yee Jun" [Incipit] [36]
Yellow Gray & Black [80]
Yellow-Sweet [33]
Yet It Is [51]
Zimzum Is Three [17; 44]
"a zone around the earth" [Incipit] [191]

III On ruth weiss (Selection)

III.1 Academic essays, theses, book chapters, etc.

Antonic, Thomas. "From the Margin of the Margin to the 'Goddess of the Beat Generation': ruth weiss in the Beat Field, or: 'It's Called Marketing, Baby.'" *Out of the Shadows: Beat Women Are Not Beaten Women.* Ed. Frida Forsgren and Michael J. Prince. Kristiansand: Portal Books, 2015. 179–199.

Antonic, Thomas. "'God's Empty Chair,' or: You Can't Dig It – Jazz & Poetry of the Beat Generation and Beat-Inspired Jazz & Poetry in Austria." *Jazz in Word: European (Non-) Fiction.* Ed. Kirsten Krick-Aigner and Marc-Oliver Schuster. Würzburg: Königshausen & Neumann, 2018. 331–346.

Belletto, Steven. "ruth weiss Keeps the Beat." [Chapter 9.2 of] *The Beats: A Literary History.* By Steven Belletto. Cambridge: Cambridge University Press, 2020. 246–251.

Carden, Mary Paniccia. "Consociation: ruth weiss's Desert Journal, For These Women Of The Beat, and Can't Stop The Beat" [Chapter 4 of] *Women Writers of the Beat Era: Autobiography and Intertextuality.* By Mary Paniccia Carden. Charlottesvillle: University of Virginia Press, 2018. 82–104.

Castelao-Gómez, Isabel, and Natalia Carbajosa Palmero. "ruth weiss – 'yo ya estaba allí': El descubrimiento de una precursora del jazz-beat." [Spanish] *Female Beatness: Mujeres, género y poesía en la generación Beat.* Valencia: PUV, 2019. 269–310.

Encarnación Pinedo, Estíbaliz. Beat & Beyond: *Memoir, Myth and Visual Arts in Women of the Beat Generation / Más Allá del "Beat": Memoria, Mito y Arte Visual en las Mujeres de la Generación Beat.* Diss. University of Murcia, 2016.
[Herein especially chapter IV.4: "Visual Arts in ruth weiss." 368–437.]

Encarnación Pinedo, Estíbaliz. "Expanded Poetry and the Beat Generation: The Case of ruth weiss" [Chapter 9 of] *The Beats: A Teaching Companion.* Ed. Nancy M. Grace. Clemson: Clemson University Press, 2021. 121–132.

Grace, Nancy M. "ruth weiss's *Desert Journal*: A Modern-Beat-Pomo Performance." *Reconstructing the Beats*, Ed. Jennie Skerl. New York: Palgrave Macmillan, 2004. 57–71.

Höfer, Hannes. "Literatur und Jazz: Zum Beispiel ruth weiss." [German] *literaturkritik.de* 8 (Aug. 2019). Online: https://literaturkritik.de/literatur-und-jazz-zum-beispiel-ruth-weiss,25865.html [accessed 2020–11]

Lee, A. Robert. "Performance Art, Art Performance." [Chapter about ruth weiss and Anne Waldeman in] *The Beats: Authorships, Legacies.* Edinburgh: Edinburgh University Press, 2019. 144–148.

Pointl, Stefanie. *"This Is My Home – This Wandering": Exile and Transnational Mobility in ruth weiss's Poetry.* MA thesis. University of Vienna, 2019.

Whaley, Preston. *Blows Like a Horn: Beat Writing, Jazz, Style, and Markets in the Transformation of U.S. Culture.* Cambridge, MA: Harvard UP, 2004.
[chapter "On the Brink", pp. 50–81, on ruth weiss and Bob Kaufman]

III.2 Interviews and conversations

Antonic, Thomas. "Vienna Never Left My Heart: A Conversation with ruth weiss, Recorded and Transcribed by Thomas Antonic." *European Beat Studies Network* 2014. Online: https://ebsn.eu/scholarship/interviews/ruth-weiss-interviewed-by-thomas-antonic/ [accessed 2020–11–01]

Brents, Walker. "A Conversation with ruth weiss." *San Francisco Reader* (Sept. 2002): 11–13.

Grace, Nancy M., and Hallie Shapiro. "Single Out." *Breaking the Rule of Cool: Interviewing and Reading Women Beat Writers*. Ed. Nancy M. Grace and Ronna C. Johnson. Jackson: UP of Mississippi, 2004. 55–80.

Grizzell, Patrick. "So There That You Are in It." *The Tule Review: A Literary Quarterly* (Sacramento) (Spring/Summer 1994): 3–8.

Janícek, Jeroným. "Člověk má věřit instinktům, I když přitom nadělá spoustu." [Czech]. *Lidové Novine* (Prague), 29 April 1998: 10.

N. N. "Four Thousand Years of Romance: An Interview with ruth weiss and Paul Blake." *A & E: Arts & Entertainment Magazine* (Mendocino, CA), May 1983: 8–13.

Knight, Brenda. "Interview – Ruth Weiss." *Women's National Book Association, San Francisco Chapter*, n.d. [2016]. Online: http://wnba-sfchapter.org/interview-ruth-weiss/ [accessed 2020–11]

Knight, Brenda. "It's 2:45, 1956: 2 People Talking." Brenda Knight and ruth weiss in Conversation. *Speak* (July–August 1998): 90–93.

Ring, Kevin. "ruth weiss: Poetry & All That Jazz." *Beat Scene* 38 (2000): 23–24. And *Beat Scene* 39 (2001): 26–28. [2 parts]

Rokko [Clemens Marschall]. "The Beat Goes On." [German] *Rokko's Adventures* 1:1 (2007): 12–17.

Vernor, Kara. "ruth weiss: Beat Poet and Bohemian Revolutionary." *Rhythmic Revue* (Arcata, CA) 10 (March 1998): 5.

III.3 Portraits, newspaper articles, reviews, etc.

Aigner, Hal. "'Surprise Voyage'. An Adventure of Words." *San Francisco Chronicle* [suppl. "Datebook"], 11 Feb. 1973: n.p.
[portrait]

Antonic, Thomas. "Die Stimme des Widerstands." *Falter* (Vienna), 21 Aug. 2018: 24–26. Online version: https://www.falter.at/zeitung/20180821/die-stimme-des-widerstands?ver=b [accessed 2020–11]
[portrait on occasion of weiss's 90th birthday]

Antonic, Thomas. "Jazz & Poetry Pioneer and Word Shaman ruth weiss, 1928–2020." *European Beat Studies Network* (Aug. 2020) Online: https://ebsn.eu/scholarship/voices/jazz-poetry-pioneer-and-word-shaman-ruth-weiss-1928–2020/ [accessed 2020–11] Shortened German version: "ruth weiss 1928–2020." *Der Standard* (Vienna), 14 Aug. 2020, [suppl. "Album"]: A6. Online version under the title "Jazz-&-Beat-Poetry-Pionierin ruth weiss gestorben": https://www.derstandard.at/story/2000119335126/ruth-weiss-19282020 [accessed 2020–11]
[obituary]

Antonic, Thomas, Dir. *One More Step West Is the Sea: ruth weiss*. Documentary film, 93 minutes, Austria 2021.

Borcich, Zida. "Poetry Matters: ruth weiss." *Real Estate Magazine* (Fort Bragg, CA), 28 June 2013: 2–4.

Brinks, Dave. "The Girl with the Green Hair: Meet the Poet ruth weiss." *Entrepôt* (New Orleans) 1:2 (Oct. 2011): 1–3.
[portrait with focus on ruth weiss in New Orleans in the 1950s]

Burton, Tony. "ruth weiss, American Beat Poet." *Lake Chapala Artists* (June 2015). Online: http://lakechapalaartists.com/?p=2311 [accessed 2020–11]
[portrait]

Caen, Herb. "In My Merry Dotmobile." *The San Francisco Chronicle*, 15 Jan. 1993: B1.
[The column in which the author famously labeled ruth weiss "Beat Generation goddess"; in Brenda Knight's *Women of the Beat Generation* on p. 241 wrongly dated Feb. 15, 1993]

Calder, Chris. "Beat Poet ruth weiss Is Riding a Wave." *The Mendocino Beacon* 120:23 (1997): 1+.

Carbajosa, Natalia. "Oralidad, jazz y poesía: ruth weiss." [Spanish] *Jot Down: Contemporary Culture Mag* (2016). Online: https://www.jotdown.es/2016/09/oralidad-jazz-poesia-ruth-weiss/ [accessed 2020–11]

Conway, Laura. "Riffs (conversation with ruth weiss)." *Optimism* (Prague) 26 (May 1998): 17–18.

Cooney, Ellen. "Shorts." *Contact II. A Bimonthly Poetry Review*, (Fall/Winter 1980): n.p.
[review of *Single Out*]

Czerny, Karin. "The Beat Goes On." [German] *Falter* (Vienna), 17 May 2002. Online: https://www.falter.at/zeitung/20020515/the-beat-goes-on/_1910460026?ver=b [accessed 2020–11–04]

Debar, Janet. "Janet Debar on ruth weiss." *Real Estate Magazine* (Fort Bragg, CA), 28 June 2013: 4.

French, Warren. "The Matriarch of the Beats." *The Kerouac Connection* 24 (Winter 1992): 25–28.
[portrait]

Guzelimian, Ara. "3 Premieres at Bing Theater." *Los Angeles Times*, 15 Dec. 1976: N.p.
[review of "Fortieth Day" from *Desert Journal*, score by Gerhard Samuel]

Hamlin, Jesse. "Weiss Riffs to the Beat of Poetry and All That Jazz." *San Francisco Chronicle* [suppl. "Datebook"], 6 Aug. 1998: E1.
[portrait]

Hernández, Hortensia. "Ruth Weiss, poeta beat." [Spanish] *Heroínas*, 24 June 2020. Online: http://www.heroinas.net/2020/06/ruth-weiss-poeta-beat.html [accessed 2020–11]
[portrait]

Huckaby, Gerry. "A Fool's Journey. Haiku Paintings by ruth weiss."*A & E: Arts & Entertainment* (Mendocino, CA) 19:4 (April 1994): 6–7.

Kikel, Rudy. "Ruth Weiss – Poet of Devastation, Demolition." *GCN* (Boston), 26. Mar. 1977: 14.
[review of *Desert Journal*]

Knight, Brenda. "ruth weiss: The Survivor (1928–)." [Chapter of] *Women of the Beat Generation: The Writers, Artists, and Muses at The Heart of a Revolution*. Berkeley: Conari Press, 1996. 241–247.

Kaczorowski, Mary Rose. "Can't Stop the Beat: Albion's ruth weiss." *Mendocino Beacon*, 15 Dec. 2017. Online: https://www.mendocinobeacon.com/2017/12/15/cant-stop-the-beat-albions-ruth-weiss/

Kerr, Mary. "Out of the Shadows: Women in the Cultural Underground-1950s/60s." *Out of the Shadows: Beat Women Are Not Beaten Women*. Ed. Frida Forsgren and Michael J. Prince. Kristiansand: Portal Books, 2015. 241–264.
[short portrait of ruth weiss on pages 256–257]

Knight, Brenda. "Return of the Prodigal Poet: ruth weiss in San Francisco Poetry Festival July 24" *The Examiner*, 22 July 2009. Online: https://www.mail-archive.com/sixties-l@googlegroups.com/msg02444.html [accessed 2020–11]

Köck, Samir H. "Die Göttin aus dem Keller. Porträt: ruth weiss war eine von ihnen und überlebte sie fast alle – die Beatniks." [German] *Die Presse* (Vienna), 24 Oct. 2006: 33.

Lavender, Bill. "For Devotees to the Grape and the Vanished American Dream." *Entrepôt* (New Orleans) 1:1 (Sep. 2011): 3.
[on the history of Jazz & Poetry in New Orleans, first paragraphs dedicated to ruth weiss]

Ljubas, Irena. "Mutti und ich gehen Hand in Hand weg und schauen nicht zurück." [German] *A Letter to the Stars: Holocaust – Die Überlebenden: Schüler schreiben Geschichte*. Ed. Andreas Kuba, Josef Neumayr, Markus Priller, and Alfred Worm. Vienna: Verein Lernen aus der Zeitgeschichte, 2005. 299–301.
[portrait with focus on ruth weiss' escape from Nazi-Germany/Austria]

Martinetti, Ron. "To GoGo for Philip: ruth weiss and Friends." *American Legends*. Online: http://americanlegends.com/authors/gogo.html [accessed 2020–11]

Neundlinger, Helmut. "Nachtvogel singt." [German] *Datum* (Vienna) 1 (2007): 64–67.
[portrait]

N. N. "ruth weiss in Memoriam." *The Beat Museum*, 1 Aug. 2020. Online: https://www.kerouac.com/ruth-weiss-in-memoriam/ [accessed 2020–11]

Peterson, Art. "When Bebop Filled the Night." *The Semaphore* 194 (Spring 2011): 1+.

Pietarinen, Fred. "ruth weiss: 'a new view of what matters'." *City Arts Monthly*, Nov. 1981: 15.
[portrait]

Ray, W. J. "An Ephemeral Re-occurence: W. J. Ray Reviews *The Brink* by Ruth Weiss." *The New Settler* (Willits, CA) 40 (1989): 43–44.

Seid, Steven. "On the Brink of Something." *UC Berkeley Art Museum and Pacific Film Archive (BAMPFA)*. Online: https://bampfa.org/page/out-of-the-vault-essay-ruth-weiss-brink [accessed 2020–11]

Seid, Steven. "The Forgotten Artists of the Belvedere Codfishery." *Open Space*, June 20, 2019. Online: https://openspace.sfmoma.org/2019/06/the-forgotten-artists-of-the-belvedere-codfishery/ [accessed 2020–11]
[on one of the filming locations of *The Brink*]

Shectman, Lydia. "Ruth Weiss: Writing into the 'Elegant Eighties'." *The San Francisco Sentinel*, 3 Nov. 1977: 4.
[portrait, preview of a reading at Ft. Mason]

Spandler Gisa. "Wo echte Hippie-Kultur überlebte: Besuch bei einer der letzten lebenden Beat-Dichterinnen in Nordkalifornien." [German] *Der Bote* (Feucht, Germany), 21 Sep. 2013: 6.
[portrait]

Spandler, Horst. "ruth weiss and the American Beat Movement of the '50s and '60s. / ruth weiss und die Beats im Amerika der 50er und 60er Jahre." [English and German versions] *No Dancing Aloud*. By ruth weiss. Vienna: Edition Exil, 2006: 213–247. [English version only also published in *Can't Stop the Beat: The Live and Words of a Beat Poet*. By ruth weiss. Studio City [Los Angeles], CA: Divine Arts, 2011. IX–XXVI; German version only also published under the title "'Timing is what matters' – Der lange Weg von ruth weiss" in *Einen Schritt weiter im Westen ist die See*. By ruth weiss. Wenzendorf: Stadtlichter Presse, 2012. 152–173.]
[biographical essay]

Spandler, Horst. "ruth weiss Comes Home: Festival in Vienna." *Beat Scene* 41 (Spring 2007). 38.

Stock, Doreen. Review of *Light* and *Desert Journal*. *Boston Gay Review*, Fall 1978: n.p.

Swoop, Dawn. "ruth weiss: Poetry & All That Jazz Vol. 1 & 2." *Beat Scene* 33 (1999): 55.

Swoop, Dawn. "No Dancing Aloud by ruth weiss (Edition Exil)." *Beat Scene* 41 (Spring 2007): 61.

Trigilio, Tony. "An Introduction: A New View of ruth weiss." *e-poets.network* (May 2003). Online: http://voices.e-poets.net/weissr/intro.shtml [accessed 2020–11]

Turner, Raymond Nat, and Zigi Lowenberg, "ruth weiss: San Francisco's Jazz & Poetry Innovator." *Jazzsteps* (San Francisco) 1:10 (2002): 8.

Wolfe, Abby. "What's Love Got to Do with It? A Women-led Tour of the Beat Generation with Eileen Kaufman, Joanna McClure and ruth weiss." *North Beach Now* (San Francisco) 11:2 (Feb. 1997): 9–10.

Zobel, Greg. "ruth weiss: Making a Deeper Groove in Europe." *Kerouac Connection* 29 (1998): 31–33.

Notes on Contributors

Thomas Antonic, Dr. phil. [PhD], Beat scholar, poet, filmmaker, musician, currently leading the research project "Transnational Literature: Austria and the Beat Generation" at the University of Vienna. Visiting Scholar at the FSI (Freeman Spogli Institute for International Studies) at Stanford University 2013, Max Kade Fellow at the Department of German Studies at the University of California, Berkeley, 2014/15. His most recent publications include the books *Amongst Nazis // Unter Nazis: William S. Burroughs in Vienna 1936/37* (bilingual edition, Engl./Germ., Moloko 2020), *Wolfgang Bauer: Werk, Leben, Nachlass, Wirkung* (German, Ritter 2018), the poetry collection *Flickering Cave Paintings of Noxious Nightbirds / Flackernde Felsbilder übler Nachtvögel* (bilingual Engl./Germ., Ritter 2017), and the CD *Fat Cat Bonfire* with his band William S. Burroughs Hurts (Moloko Plus, 2019). He is the director of the feature length documentary film *ruth weiss: One More Step West Is the Sea* which premiered in 2021. A biography about ruth weiss is in process.

Steven Arnold was an American multidisciplinary artist, filmmaker, photographer, painter, illustrator, set and costume designer. Born in 1943 in Oakland, he studied at San Francisco Art Institute and at the École des Beaux-Arts in Paris in the early 1960s. In the late 1960s he started to make short films, many starring ruth weiss. His sole feature length film *Luminous Procuress* (1972), with weiss in a main role, won the New Director's award at the San Francisco International Film Festival. His films were also screened at the Cannes Directors' Fortnight, the Chicago International Film Festival, and the Toronto International Film Festival. Salvador Dalí was so impressed with *Luminous Procuress* that he arranged a private screening at the St. Regis Hotel, to which he invited New York's elite, including Andy Warhol, who also praised the film's genius. He established a Los Angeles photography studio and west coast salon, and from 1982 to 1989 he found his niche, designing and shooting tableau-vivants for four books. He left thousands of living tableau photographs and negatives unpublished and died in 1994.

Caroline Cooley Crawford has been a music historian at the Bancroft Library, University of California/Berkeley for thirty years. A native Californian, she has degrees from Stanford University, the University of Geneva and the Royal College of Musicians, London. Since the 1980s she has reviewed music performances for Bay City News Wire Service. She has conducted more than thirty oral histories in music, including subjects in jazz and blues (Dave Brubeck, John Handy, Jimmy McCracklin) and American composers (Andrew Imbrie, Henry Brant, Pauline Oliveros, Joaquin Nin-Culmell). Those with multiple narrators include: *Doctor Atomic: The Making of an American Opera*; *David Harrington and the Kronos Quartet: Musician without Borders*; *Ali Akbar Khan and the North Indian Classical Music Tradition*; and *Kurt Herbert Adler and the San Francisco Opera*. She produced and directed a blues documentary entitled *Jimmy Sings the Blues* and recently published *The California Blues: A Musical Journey from the South to the West Coast*.

Neeli Cherkovski is an American poet and memoirist, who has resided since 1975 in San Francisco. He has written biographies of Lawrence Ferlinghetti, and Charles Bukowski, with whom he co-edited the Los Angeles zine *Laugh Literary and Man the Humping Guns*. He is also the author of *Whitman's Wild Children* (1989), a collection of essays about twelve poets, among them Allen Ginsberg, Michael McClure, Bob Kaufman, Gregory Corso, and Philip Lamantia. He taught literature and philosophy at the New College of California in San Francisco. His body of poetry includes *Animal* (1996), *Elegy for Bob Kaufman* (1996), *Leaning Against Time* (2004), and *Elegy for My Beat* Generation (2018). He was awarded the 15th Annual PEN Oakland/Josephine Miles Literary Award in 2005.

Janet DeBar was born and raised in the northern panhandle of West Virginia. She was educated at the College of Wooster in Ohio where she tried to resist the influence of Presbyterianism and at Stanford in California where she similarly tried to resist the influence of Yvor Winters. She married and was supported by her husband in the old, outmoded style, leaving her free to pursue a rather goliardic existence, auditing Latin classes at UC Berkeley and learning hand press printing from William Everson (a.k.a. Brother Antoninus) at UC Santa Cruz. She and her husband have lived on the North Coast of California for over twenty years during which time she has organized several poetry reading series and participated in many organized by others. Her poems have appeared in *Wood, Water, Air and Fire: The Anthology of Mendocino Women Poets*, *The Redwood Coast Review* and in the *Café Review*.

Estíbaliz Encarnación-Pinedo holds a PhD in postwar American Literature from the University of Murcia (Spain) and is currently a lecturer and researcher in the Department of Modern Languages at Polytechnic University of Cartagena, Murcia. Her research focuses on gender and feminism in postwar and avant-garde American poetry. She is also a member of the "Queer Temporalities" research group (University of Murcia). Her latest publications include "Expanded Poetry and the Beat Generation: The Case of ruth weiss" in *The Beats: A Teaching Companion*, ed. Nancy M. Grace (2021, Clemson University Press), "On Webbed Monsters, Revolutionary Activists and Plutonium Glow: Eco-crisis in Diane di Prima and Anne Waldman" (2020, *Humanities*) and "Shifting the Mythic Discourse: Ambiguity and Destabilization in Joanne Kyger's *The Tapestry and the Web*" (2020, *Amaltea*).

Agneta Falk was born in Stockholm, Sweden. She is a poet, visual artist, editor and translator. In her twenties she moved to England, and in 1998 she moved to San Francisco. A former educator and co-director of a literature development agency, she has organized and participated in many international poetry festivals. Her poetry is translated into many languages and she exhibits her art internationally. Her latest book, *Heart Muscle* was published in 2009 by Multimedia Edizioni, Italy. She is a member of Revolutionary Poets Brigade and recently took over the Life Worms Gallery in San Francisco.

Frida Forsgren is an art historian working on the Beat artists in San Francisco. She has published the books *San Francisco Beat Art in Norway* (2008) and *Beat Lives* (2013), and is the co-editor of the volumes *Out of the Shadows: Beat Women Are not Beaten Women* (2015) and *Multimodality and Aesthetics* (2019). She has made a full catalogue of the Wennesland Collection, the world's largest collection of San Francisco Beat Art. At the Unversity of Agder, Norway, she teaches the interdisciplinary course on Beat Culture.

S.A. Griffin lives, loves and writes in Los Angeles.

Benjamin J. Heal is currently assistant professor of American literature at National Chung Cheng University in Taiwan, and has presented conference papers and guest lectures on his work internationally. His work includes essays in the collections *Do You Bowles? The Next Generation* (Brill, 2014) and *The Routledge Handbook of International Beat Literature* (2018), and an article in the 2016 European Beat Studies Network (EBSN) special issue of *CLCWeb: Comparative Literature and Culture*. A current elected board member of the EBSN, Benjamin has organized literary events and was a recipient of a Harry Ransom Center fellowship in 2014. His wider research interests include Chinese literature, transnationalism, and the works of James Leo Herlihy.

Jack Hirschman is an American poet and social activist who has written more than 50 volumes of poetry and essays. He received his PhD from Indiana University in 1961 and taught at the University of California in Los Angeles until being fired after encouraging his students to resist the draft for the Vietnam War. After that he moved to San Francisco where he still lives. Hirschman is also a painter and collagist, and has translated over two dozen books from German, French, Spanish, Italian, Russian, Albanian, and Greek. Among his many volumes of poetry are *A Correspondence of Americans* (1960), *Black Alephs* (1969), *Lyripol* (1976), *The Bottom Line* (1988), and *Endless Threshold* (Curbstone, 1992). His magnum opus *The Arcanes* was published in 2006. In the same year he was appointed Poet Laureate of San Francisco.

Hannes Höfer, Dr. phil., is research associate at the Institut für Germanistische Literaturwissenschaft at Friedrich-Schiller-Universität Jena. He is the author of *Deutscher Universalismus: Zur mythologisierenden Konstruktion des Nationalen in der Literatur des 18. Jahrhunderts* (2015). He has published articles on Goethe in Naples, Arno Schmidt in post-war Germany, Mack the Knife in the USA, and jazz in the work of ruth weiss: "Literatur und Jazz: Zum Beispiel ruth weiss" (2019) was published in the online journal *literaturkritik.de*. In 2018/19 Hannes Höfer has been Volkswagen Foundation research fellow at the Washington University in St. Louis. He is currently writing a book on jazz in German-language fiction.

Mary Norbert Körte is a poet and San Francisco Bay Area native who lives in Willits, Northern California. She became a nun at the St. Rose Convent in San Francisco in 1952 and earned a master's degree in Silver Latin. In 1965 she became part of the Beat Movement, left the convent in 1968 and became a secretary of the psychological department of the University in California. In 1972 she was nominated for the National Endowment award in 1972 and moved to Northern California where she became part of the Save the Redwoods movement and taught writing in the Poetry in Schools project and at the college on a Native American reservation near her home. She published numerous poetry collections and broadsides, including *The Beginning of Life* (1967), *The Midnight Bridge* (1970), *Mammals of Delight* (1978), and *Throwing Firecrackers out the Window While the Ex-Husband Drives By* (1991).

A. Robert Lee, formerly of the University of Kent, UK, was Professor of American Literature at Nihon University, Tokyo 1996–2011. He has held visiting professorships at Northwestern University, the University of Colorado and Berkeley. His forty or so book publications include *Designs of Blackness: Mappings in the Literature and Culture of Afro-America* (1998) and *Multicultural American Literature: Comparative Black, Native, Latino/a and Asian American Fictions* (2003), which won the 2004 American Book Award. His work on Beat authorship includes the edited collection *The Beat Generation Writers* (1996), *Modern American Counter Writing: Beats, Outsiders, Ethnics* (2010), *The Routledge Handbook of International Beat Literature* (2018), and most recently, *The Beats: Authorships and Legacies* (2019) and *The Joan Anderson Letter: The Holy Grail of the Beat Generation* (2020).

Polina Mackay is the Vice President of the European Beat Studies Network and Association and Associate Professor of English and Head of the Department of Languages and Literature at the University of Nicosia. She is the author of *The Aesthetics, Gender and Feminism of the Beat Women* and the co-editor of *Global Beats* (with Oliver Harris), *The Beat Generation and Europe* (with Chad Weidner), *The Cambridge Companion to H.D.* (with Nephie Christodoulides), *Kathy Acker and Transnationalism* (with Kathryn Nicol) and *Authorship in Context* (with Kyriaki Hadjiafxendi). Her work also appears in *The Routledge Handbook of International Beat Literature* and *The Cambridge Companion to the Beats*. Her work on the Beats has also been translated into French.

Lars Movin, writer and filmmaker, specializing in the Beat Generation and avant-garde art and film. As an author and/or editor he has published over thirty books since 1990, many of them on subjects such as video art, American avant-garde film, and the New York downtown-scene; but also monographs or biographies on cultural figures like Captain Beefheart, Jørgen Leth, Gerard Malanga and Torben Ulrich. One of his main works is a 600+ page study of the Beat writers published in 2008, and in later years he has written books on Allen Ginsberg (2019) and William S. Burroughs (2021). Among his circa twenty documentaries, some of them in collaboration with Steen Møller Rasmussen, one can find *The Misfits – 30 Years of Fluxus* (1993), *Connections: Ray Johnson On-Line* (2001), and *Words of Advice – William S. Burroughs on the Road* (2007).

Peggy Pacini is Associate Professor of American literature at CY Cergy Paris Université. Her interests in scholarship include Beat studies, live poetry, cultural production and communal identity. Her publications include *Lettres Choisies (1943–1997)*. French translation of *The Letters of Allen Ginsberg* (2013), "City Lights and the Emergence of Beat Poetry: How *Howl and Other Poems* Redefined Poetic and Cultural Boundaries in the mid-1950s" (2017), "Êtes-Vous Beat? Contemporary French Beat Writing" (2018), "Je pense à Jack Kerouac, le père que nous n'avons jamais trouvé (mythe, nostalgie et réalité)" (2018) and "HOWL, construction du mythe et avènement de la communauté?" (2020).

Paul Pechmann, literary critic, lecturer and researcher, since 2017 research assistant ("worker") at the University of Vienna (Department for German, project "Transnational Literature: Austria and the Beat Generation"); formerly researcher for the project "The International Reception of the 'Grazer Gruppe'" at the University of Graz / Franz Nabl Institut für Literaturforschung (Austria), and guest lecturer at the University of Shkodra (Albania). Since 2008 publishing editor for Ritter publishing house (Klagenfurt, Graz, Vienna, www.ritterbooks.com). Author and (co-)editor of numerous publications on Austrian post-war avantgarde writers (Wolfgang Bauer, Elfriede Jelinek, Gert Jonke, Gerhard Rühm, Werner Schwab and others).

Stefanie Pointl is an independent scholar of 20th century American literature with a longtime interest in the Beat Generation and transnational writing. She holds a master's degree in Anglophone Literatures and Cultures from the University of Vienna, having previously studied in

Manchester and Salzburg, where she completed an undergraduate degree. For her master's thesis "'This Is My Home – This Wandering': Exile and Transnational Mobility in ruth weiss's Poetry" (2019) she received the Fulbright Prize in American Studies. She is currently working on a monograph on the role of mobility and transnational connections in the work of ruth weiss.

Steve Seid was a media curator at the Berkeley Art Museum/Pacific Film Archive for twenty-five years. During that time, he presented almost a thousand public programs, featuring experimental media, forgotten film genres, and a sampling of international cinema. He also helped build the PFA's collection, particularly video art and personal cinemas from the Bay Area. Notable would be the restoration of Steven Arnold's *Luminous Procuress* (1971) and poet ruth weiss's only film effort, *The Brink* (1961). Seid has also been involved with several publications, most importantly, *Radical Light: Alternative Film & Video in the San Francisco Bay Area, 1945–2000*, co-edited with Steve Anker and Kathy Geritz; *Ant Farm 1968–1978*, co-edited with Constance Lewallen; and his recent solo effort, *Media Burn: Ant Farm and the Making of An Image*.

Tate Swindell is a poet, painter, photographer and archivist who likes to talk with bees. Co-editor of the *Collected Poems of Bob Kaufman* (City Lights 2019) and *On Valencia Street: Poems and Ephemera by Jack Micheline* (Lithic Press 2019). Tate runs the San Francisco based Unrequited Records which specializes in vinyl records from Beat Generation poets. Artist releases include Gregory Corso, Herbert Huncke, Bob Kaufman, Jack Micheline and Harold Norse. Tate and his brother Todd were part of the early medical cannabis movement in San Francisco. In a collective, first as the radical Queer protest group ACT UP! then as the Market Street Cooperative, they operated a dispensary for 15 years until shut down by federal government repression.

Anne Waldman has been a prolific and active poet and performer for many years, creating radical new hybrid forms for the long poem, both serial and narrative. She is the author of the magnum opus *The Iovis Trilogy: Colors in the Mechanism of Concealment* (2011) which won the PEN Center Literary Award for poetry in 2012. She has been deemed a "countercultural giant" by *Publishers Weekly*, and in 2015 was presented with a lifetime achievement award from the Before Columbus Foundation. She was one of the co-founders of the Poetry Project at St. Mark's Church-in-the-Bowery in New York City and its director for a number of years, going on to found and direct the Jack Kerouac School of Disembodied Poetics at Naropa University in Boulder, Colorado, with Allen Ginsberg and Diane di Prima in 1974, where she has continued to work as Distinguished Professor of Poetics and Artistic Director of its Summer Writing Program. She is the editor of *The Beat Book: Writings from the Beat* Generation. She makes her homes in New York City and Boulder, and reads, performs and lectures all over the world.

Stefan Weber studied German and Philosophy in Bern, Switzerland. From 1984 onwards he worked as sound designer at the Schauspielhaus Zurich theater and collaborated with Bertolt Brecht disciple Benno Besson in Switzerland, France, Italy, and Germany. After that he worked as a freelance theater director in Switzerland. In 1998 he moved to Vienna and short after met ruth weiss while she was in the city to lecture at the Vienna Poetry School (Schule für Dichtung), which was the beginning of a friendship that lasted for many years. Since

1998 he directed and produced plays in several Austrian theaters, since 2005 he is a freelance writer, director, and sound composer for radio features and radio plays of the Austrian National Broadcast (ORF).

Chad Weidner is an American/Belgian humanist and writer, and works as Head Tutor at University College Roosevelt, an undergraduate faculty of Utrecht University, in the Netherlands. He teaches liberal arts, English, and film. He was scholar in residence at the University of Colorado and later at Colorado State University. He was previously Senior Research Fellow in the Environmental Humanities at Utrecht University, where he worked with Rosi Braidotti. His PhD was written on ecocriticism at Ghent University (Belgium), and he got his MA from the University of Nebraska. He publishes on William Burroughs, the Beat Generation, ecocriticism, and film. His books include *The Green Ghost: William Burroughs and the Ecological Mind* (Southern Illinois UP, 2016), and *Fractured Ecologies* (EyeCorner P, 2020).

John Wieners (1934–2002) was an American poet who was born in Milton, Massachusetts, and earned his B.A. at Boston College. In 1955 and 1956 he studied under Charles Olson and Robert Duncan at the Black Mountain College and joined the artist community of North Beach, San Francisco, in 1957. His most famous poetry collection *The Hotel Wentley Poems* was published in 1958. In 1960 he moved back to the East Coast where he lived with Herbert Huncke. He was a Guggenheim Fellow in 1985. He published poetry books such as *Asylum Poems* (1970), *Nerves* (1970), *Behind the State Capitol or Cincinnati Pike* (1975), and *The Journal of John Wieners is to be called 707 Scott Street for Billie Holliday 1959* (1996). His *Selected Poems 1958–1984*, edited by Raymond Foye, were published at Black Sparrow Press in 1986.

A.D. Winans is a native San Francisco poet and writer. He is the author of over fifty books, including *North Beach Poems*, *North Beach Revisited*, and *This Land Is Not My Land*, which won a 2006 PEN Oakland Josephine Miles Award for excellence in literature. Recent books include *Billie Holiday, Me and the Blues*, *No Room for Buddha*, *Love-Zero*, and *San Francisco Poems*.

Pete Winslow (1934 – 1972) was a surrealist poet associated with the Beat Generation. His last volume of poems, *Daisy in the Memory of a Shark*, was published posthumously by City Lights Books in 1973. He graduated from the University of Washington in journalism in 1956. Best known for his poetry, Pete Winslow was also a newspaper columnist, essayist & novelist.

Index

Adorno, Theodor W., 35
Amran, David, 41
Anger, Kenneth, 180
Antonic, Thomas
– *One More Step West Is the Sea* (documentary), XIII, 120, 206
Arnold, Steven, XXII, 45, 68, 180, 195, 197–202, 214, 220
– *Liberation of he Mannique Mechanique, The*, 195
– *Luminous Procuress*, XXII, 197–202
– *Messages, Messages*, 195
– *Various Incantations of a Tibetan Seamstress*, 195
Ayler, Albert, 42

Baillie, Bruce, 192
Baraka, Amiri
– *New Music – New Poetry*, 115
Baudrillard, Jean
– *America*, 96
Beattie, Paul, 45, 160, 161, 179–180, 183, 190, 192–193
Behrend, Joachim Ernst, 115
Benn, Gottfried, 115
Bergman, Ingmar, 190
Blake, Paul, XI, 45, 105, 127, 160, 161, 215, 216, 219, 220
Blyth, Robert L., 43
Bowen, Michael, 175
Bowles, Paul
– "Baptism of Solitude", 46
Brady Tucker, Seth
– "The Road to Baghdad", 91
Brakhage, Stan, 45, 180, 190, 192
Brötzmann, Peter, 42
Brown, Joan, 162, 163, 191
Burroughs, William S., XIII, 42–44, 161
– *Cities of the Red Night*, 109
– *Minutes to Go*, 109
– *The Place of Dead Roads*, 109
– *The Western Lands*, 108–109

Caen, Herb, XVI, 100, 145
Cage, John, 174
Carlson, Wil, 160
Cassady, Neal, 37, 191
Cocteau, Jean, 190
Coleman, Ornette, 120
Collins, Jess, 190
Coltrane, John, 9
Conner, Bruce, 163, 165, 191
Corso, Gregory, XIII
– "Bomb", 108
Crane, Stephen
– "I Walked in A Desert", 88–89

Davis, Hal, XI, 9, 114, 120–122, 157, 205
Dawson, Daney, 113, 118–119, 124
de Beauvoir, Simone, 37
DeFeo, Jay, 165, 175, 191
Deleuze, Gilles, 42
Deren, Maya, 180
di Prima, Diane, XIII, 68, 70, 174
– *Memoirs of a Beatnik*, 59
– *Recollections of my Life as a Woman*, 70 [fn2]
Dolphy, Eric, 120
Duchamp, Marcel, 162
Duncan, Robert, 190
Dylan, Bob
– "The Times They Are A-Changin'", 38

Eliot, T.S., 81
– *Four Quartets*, 90
– *The Waste Land*, 90

Fellini, Federico, 190
Fenton, Elyse
– "Word from the Front", 91
Ferlinghetti, Lawrence, XIII, 113, 127, 128, 147, 161
Foster, Richard, 190
Frank, Robert, and Alfred Leslie
– *Pull My Daisy*, XXII, 180, 190

Frost, Robert, 81
- "Desert Places", 89

Garrison, Jimmy, 9
Gerdes, Ingeborg, XXII
Gillespie, Dizzie, 122
Ginsberg, Allen, XIII, XV, 36, 43, 47, 101, 107, 127, 128, 141, 161, 174
- "Howl", 71, 164
Gleason, Madeline, 160, 190, 220,
- "Once and Upon", 81, 92
Gysin, Brion, 42, 44

Hahn, Marianne, 77 [fn8], 160
Hedrick, Wally, 191
Hegel, Georg Friedrich Wilhelm, 42
Heidegger, Martin, 108
Heine, Heinrich, 115
Herms, George, 9, 164, 175
Hirschman, Jack, XVI
Holiday, Billie, XIII
Hofmann, Hans, 162
Hughes, Langston, 113
- "Weary Blues", 39
Hussein, Sadam, 91

Ionesco, Eugéne, 212

Jacobs, Ken, 180
Jandl, Ernst, 114–115
Jepson, Warner, 184
John, Elsie, 78
Johnson, Jay Jay, 115
Johnson, Joyce, 69
Jones, Elvin, 9
Jones, Hettie, 70
- *Drive*, 59

Kandel, Lenore, 127, 141
Kaufman, Bob, XII, 25, 40, 160, 191, 214
- "Bagel Shop Jazz", 40, 68
- *Beatitude*, 160, 212
Kennedy, John F., 40
Kerouac, Jack, XII–XIII, 36–37, 40–43, 101, 107–108, 113, 161, 165, 165 [fn11], 174, 180, 191, 208, 212
- *Book of Haikus*, 36

- "Essentials of Spontaneous Prose", 165
- *On the Road*, XV, 36, 39–40, 69
Kerouac, Jan, XIII
Kahlo, Frida, 162
Kästner, Erich, 219
Khan, Yasmine, 92
- "The Desert Night", 90
Knight, Brenda, 37, 100
- *Women of the Beat Generation*, XVI–XVII, 36
Körte, Mary Norbert, 15
Kurosawa, Akira, 190
Kyger, Joanne, XIII, 70, 100, 174

Lamantia, Philip, 191, 217
Lawyer, Lori, 183, 185–187
Leary, Timothy, 127
Losey, Joseph
- *The Boy with Green Hair*, 35
Loy, Mina
- "Mexican Desert", 90
Lynch, David
- *Blue Velvet*, 35
- *Dune*, 35

Maclaine, Christopher
- *The End*, 180
Mailer, Norman, 73, 103 [fn5]
Mandrake, Mona [a.k.a. Michael Shain], 219
Marin, Sutter, 37, 160–161, 183, 185, 188, 220
Martin, Fred, 164, 175
Masina, Giulietta, 31
Matthews, Benfaral, 9–10, 113, 116–119, 124
McCall, Steve
- *New Music – New Poetry*, 115
McClure, Michael, 101, 127, 128, 141, 174, 191
McCoy Tyner, Alfred, 9
McKeever, Anne, 160, 217
Mead, Taylor, 180, 213–214
Mekas, Jonas, 192
Menken, Marie, 180
Micheline, Jack, 25, 219
Miles, Josephine, 92
- "Desert", 89
Minger, Jack, 160

Monroe, Harriet, 92
– "A Garden in the Desert", 90
Moore, Brew, 127
Mulligan, David, 120
Murray, David
– *New Music – New Poetry*, 115

Nelson, Robert, 195
Nelson, Sonny, 160

O'Connor, Doug, 113, 120–122, 157
Osborne, Sandra, 81
– "Iraqi Desert", 91

Piaf, Édith, 31, 219
Poe, Edgar Allen
– "The Pit and the Pendulum", 86
– "The Raven", 46, 116
Pommy Vega, Janine, XIII
– *Tracking the Serpent*, 59
Powell, Bud, 155

Rice, Ron
– *The Flower Thief*, XXII, 180, 190, 213–214
Rivera, Diego, 162
Reinhardt, Ad, 162
Renoir, Jean, 179, 190
Reverdy, Pierre, 113
Rexroth, Kenneth, 113, 127, 161
Reynolds, Malvina
– "Little Boxes", 191
Romero, Elias, 76
Romus, Rent, 113, 120–123
Rothko, Mark, 162
Runyon, Scott, 67, 200, 201, 204

Samuel, Gerhard, 155
Schoen, Karl, 120 [fn3]
Seeger, Pete, 191
Shelly, Percy, 81, 88, 90, 92
– "Ozymandias", 88
Smith, Stuff, 122
Snyder, Gary, XIII, 43, 101, 127, 128, 141, 174
Spencer, Bill, 183, 184, 192
Spencer, Fumi, 160
Spohn, Clay, 162
Stauffacher, Frank, 190

Stein, Gertrude, 93
Still, Clyfford, 162, 175
Stockwell, Dean, 35
Strand, Chick, 192
Suzuki, D.T., 174
Suzuki, Shunryu, 174, 175

Tarlow, Aya, 160
Truffaut, François, 190
Tzara, Tristan, 41, 105

Varda, Jean, 162

Waldman, Anne, XIII, 68, 70
– "Feminafesto", 67–68, 73, 75
– "Hag of Beare (Caillech Berri)", 68 [fn1]
– *The Iovis Trilogy*, 68 [fn1]
Webster, Ben, 127
weiss, ruth
– *61st Year to Heaven, The*, 213, 219
– "1967", 128, 130, 133–136, 143–144
– "2009", 74
– "Africa", 74, 76
– "As the Wheel Turns", 60 [fn5]
– *B Natural*, 213
– *Banzai!*, XXI, 159, 166, 171–174
– "beast – be a saint", 128, 130, 134–142
– "Big Sur", 53
– *Blue in Green,* XXI, 38, 46, 67–68, 99, 102, 104, 179, 184, 212, 214
– *Brink, The* (film), XII, XVII, XXII, 9, 36, 44, 68, 72, 157, 161, 179–193, 212, 214, 219
– "Brink, The" (poem), 72, 184, 188–189, 192
– "Bypass Linz", 120–121
– *Can't Stop the Beat*, XII, 40 [fn3], 45, 51, 53, 57–59, 74, 100, 147, 153, 158, 160, 212
– "Chopsticks", 184, 212
– "Cockie with Earth Paintings", 38, 42
– "Compass", 53, 212
– *Desert Journal*, XII, 3, 44, 46, 61, 64, 68, 74–78, 81–99, 102, 105–106, 108, 116, 120, 123, 155, 161, 206, 214, 219
– "Earth Paintings", XII, 184
– *Figs*, 68, 71–72, 184, 212

- *Fool's Journey, A* (book), XII, XIX, 130, 166, 206
- "Fool's Journey", A (haiku series), XXI, 159, 166–169, 172, 174
- "For Bobby Kaufman", 40
- *For These Women of the Beat*, 74 [fn5], 78, 215
- "Forty-One Dragon Steps", 41–42
- *Full Circle* (book), XVIII
- "Full Circle" (poem), 77
- *Gallery of Women*, 37–38, 42, 47, 161, 184, 212, 215
- "Her Number Was Not Called", 62, 64
- "I Always Thought You Black", 40 [fn3], 57, 59–60, 63, 120, 122–124
- "In a Japanese Tree Garden", 102–103
- *Las Cuevas de Albion*, 179
- "Light", 76–78, 184
- *Light and Other Poems*, XII, 68, 74, 184, 214
- *M & M*, 212, 213
- *make waves*, 129, 130, 134, 138, 139, 144
- *New View of Matter, A*, 157
- "No Credit", 206
- *No Dancing Aloud* (book), 62, 212
- *No Dancing Aloud* (play), 212
- "One More Step West Is the Sea", 41–42, 44, 54, 56, 63, 70, 209
- *One Knight and One Day*, 212
- *Parallel Planet for People and Places, A*, XIII, 74
- "Promise Kept", A, 63
- *ruth weiss – 2018 Live at The Beat Museum and Other Recordings from the Archive*, 219
- "Sena", 39
- *Single Out* (book), XII, 148, 179, 212
- "Single Out" (poem), 55, 63, 77, 209, 219
- "Something Current", 73
- *South Pacific*, XXI, 74, 77 [fn8], 99, 102, 161, 212
- "speak for yourself", 128, 130, 143–144
- *Steps*, 41, 54, 70, 209, 212
- *Stop That Flower*, 213, 214
- *Thirteenth Witch, The*, 72 [fn3], 212
- *Tomboy at Boarding School*, 157
- "Torch Song for Prometheus", 60 [fn5], 212
- "Words No Words for a Dance Fumi", 184

Weitsman, Mel (a.k.a. Sojun Roshi), XII, 38, 160, 175, 175 [fn16], 181, 184, 190, 211, 212, 217, 219
Wenders, Wim
- *Paris, Texas*, 35, 96
Whalen, Philip, XV, 43, 174
Wieners, John
- *Hotel Wentley Poems*, XV
Winans, A.D., 10
Wong, Kaisik, 202
Wordsworth, William, 81, 88, 90, 92–93
- *The Prelude*, 87

Yaski, Richard, 161
Yeats, William Butler, 41

Zimmerman, Vernon
- *Lemon Hearts*, 180
Zoller, Attila, 115

www.ingramcontent.com/pod-product-compliance
Lightning Source LLC
Chambersburg PA
CBHW071736150426
43191CB00010B/1600